Autistic Community and th
                    Movement

Steven K. Kapp
Editor

Autistic Community and the Neurodiversity Movement

Stories from the Frontline

Editor
Steven K. Kapp
College of Social Sciences
and International Studies
University of Exeter
Exeter, UK

Department of Psychology
University of Portsmouth
Portsmouth, UK

ISBN 978-981-13-8439-4   ISBN 978-981-13-8437-0 (eBook)
https://doi.org/10.1007/978-981-13-8437-0

© The Editor(s) (if applicable) and The Author(s) 2020. This book is an open access publication.

Open Access This book is licensed under the terms of the Creative Commons Attribution 4.0 International License (http://creativecommons.org/licenses/by/4.0/), which permits use, sharing, adaptation, distribution and reproduction in any medium or format, as long as you give appropriate credit to the original author(s) and the source, provide a link to the Creative Commons license and indicate if changes were made.

The images or other third party material in this book are included in the book's Creative Commons license, unless indicated otherwise in a credit line to the material. If material is not included in the book's Creative Commons license and your intended use is not permitted by statutory regulation or exceeds the permitted use, you will need to obtain permission directly from the copyright holder.

The use of general descriptive names, registered names, trademarks, service marks, etc. in this publication does not imply, even in the absence of a specific statement, that such names are exempt from the relevant protective laws and regulations and therefore free for general use.

The publisher, the authors and the editors are safe to assume that the advice and information in this book are believed to be true and accurate at the date of publication. Neither the publisher nor the authors or the editors give a warranty, expressed or implied, with respect to the material contained herein or for any errors or omissions that may have been made. The publisher remains neutral with regard to jurisdictional claims in published maps and institutional affiliations.

Cover image: Ryan Smoluk

This Palgrave Macmillan imprint is published by the registered company Springer Nature Singapore Pte Ltd.
The registered company address is: 152 Beach Road, #21-01/04 Gateway East, Singapore 189721, Singapore

Foreword

This book describes some of the key actions that have defined the autism rights branch of the neurodiversity movement since it organized into a unique community over 20 years ago. The actions covered are legendary in the autism community and range from "The Autistic Genocide Clock" through to the "Institute for the Study of the Neurologically Typical", and famous pieces of work like "Don't Mourn for Us".

These acts have forged new thinking on autism and established the neurodiversity movement as a key force in promoting social change for autistic people. It is primarily autistic activists who have been at the vanguard of the neurodiversity movement. All but two of the 21 contributors to this volume identify as autistic. The collection describes the biographies and rationale of key activists **in their own words**, thus the motto of disability rights activism "nothing about us without us" is a guiding tenet for the book. The phrase (and this volume) are rooted in the concept of *standpoint epistemology*. A standpoint position claims that authority over knowledge is created through direct experience of a condition or situation. Standpoint epistemology is related to the idea of lay expertise, which is discussed extensively in the sociological literature. So, the book values the experience of autistic people as a source of knowledge about their own plight.

The volume acknowledges that individual contributions are shaped by contributors' political and social experience as well as their lived understanding of autism. Standpoint theory suggests inequalities foster particular standpoints, and that the perspectives of marginalized and oppressed groups can generate a fairer account of the world. Individuals from such groups are in a distinctive position to call out forms of behavior and practices of the dominant group, hopefully leading to social change. This collection illustrates the perspective of each contributor's unique voice. But together, the chapters powerfully illustrate the sense of a group with a shared point of view, united in a common movement.

Enormous credit goes to the Editor, Steven K. Kapp, who bought this volume together and was able to simultaneously command the trust of the autism activist community and the respect of the academic community—not an easy balance to get right!

The hope is this book will provide a reference text for readers interested in the history and ideas of the neurodiversity movement and how these ideas have shaped production of expert and especially lay knowledge about autism. However, the neurodiversity movement has been problematized by both parents and academics for being unrepresentative and divisive, and the book also addresses some of these critiques.

The target academic audience is primarily undergraduates and scholars in sociology, history of medicine, and psychiatry. This collection of activists' stories should act as a reference text useful as a source for further academic debate and analysis. Another important set of readers are parents wishing to learn more, and of course autistic persons themselves. Our generous funder, the UK's Wellcome Trust, has supported the publication of this volume on an open access platform to make it available for free online.

The book is not a complex analysis or a "celebratory" piece; instead it offers raw first-person accounts, relating how and why activists have contributed. It aims to preserve and document the stories of some of the original activists whose voices helped shape and inspire the fledgling neurodiversity movement.

Enjoy!

Exeter, UK   Ginny Russell

Preface

This book has emerged from a postdoctoral research fellowship within sociology in the U.K. as part of the Wellcome Trust-funded project *Exploring Diagnosis: Autism and the Neurodiversity Movement*, which includes academic engagement with the movement as well as critical analysis of its position. The following chapters provide an overview of the neurodiversity movement, describing the key actions of autistic activists in the movement between 1992 and the present day, in their own words. Although previous books have provided coverage of the history of autism inclusive of the neurodiversity movement as narrated by a journalist or researcher [10, 12] or featured anthologies of contributors from autistic people published within the movement [1, 3, 9], this edited collection provides the first history of the movement from firsthand accounts of members of the autistic community and both autistic and non-autistic parent movement activists. Following my introduction to the movement and contributions, the book contains 19 chapters by 21 authors organized into parts about the forming of the autistic community and neurodiversity movement, progress in their influence on the broader autism community and field, and their possible threshold of the advocacy establishment. This is followed by a description of some critiques of the movement, and I follow with the conclusion.

The *Exploring Diagnosis* project research inquires into how diagnosis catalyzes mobilization, focusing on autistic adults and the neurodiversity movement (NDM). The project aimed to assess what the understandings of neurodiversity are among autistic adults. Ginny Russell, the project lead, and I conducted a thematic analysis [2] of autistic adults' responses to the question "What is neurodiversity in your own words" from a study I co-led with Kristen Gillespie Lynch et al. [4]. Russell and I found that the data largely mapped onto definitions autistic adults in the movement have given [5, 11] defining the NDM as encompassing both human biological differences in cognition, brains, and genes while also serving as an activist tool for change toward acceptance and inclusion of autistic and other *neurodivergent* people. Responses described neurodivergent people as socially oppressed and stigmatized, yet possessing valuable differences that should not be cured but instead supported with rights and accommodations. Considering that research participants knew of but did not necessarily subscribe to the movement, the consensus that emerged suggested clarity about the meaning of (the) neurodiversity (movement) as understood by aware autistic adults. Indeed, the descriptions of these terms from contributors to this book who chose to put them forward, aligned well with those in both the popular literature and aforementioned analysis.

As contributors (including myself) make clear, we think autism involves strengths and weaknesses that amount to both a difference *and* a disability, and do not consider autism an advantage overall but see autistic people as socially disadvantaged. No one emphasized strengths overall or highlighted particular strengths, but we take a "big-tent" approach to autism that recognizes no one's worth depends on having particular talents or abilities. The *Exploring Diagnosis* project corroborated this further by interviewing autistic adults in the UK about the strengths participants think autism confers, which almost everyone identified, and yet they mentioned moderating influences that could make these traits function as challenges [8].

Conversely, the book most addresses the question of what political activities the NDM has taken part in, and how these have challenged the notions of diagnosis and intervention for autism. The contributors illustrate notable examples of the manifestos (both ideological and as applied to policy), mailing lists, websites, conferences, issue campaigns, academic projects and journals, books, organizations, and advisory roles to parent- and professional-led bodies that constitute some of the range of the neurodiversity movement's political activities. These actions have had widespread impacts toward an emerging view among families, practitioners, researchers, and the public on autism as *both* a difference and disability.

Positionality Brought to the Book

I am both an autism scholar and an autistic neurodiversity activist, so while I seek to maintain high standards of rigor and fairness in editing this collection, it may reflect this positionality. In 2007 Scott Robertson, the co-founder and then Vice President of the Autistic Self Advocacy Network, reached out to me privately on Facebook and undertook a mentorship role in which he introduced me to the neurodiversity and disability rights movements. At that time, as an undergraduate, I lived in my hometown of Los Angeles in the world's region (Southern California) most dominated by the mainstream medical and alternative medical autism establishment. Groups like the Center for Autism and Related Disorders (CARD) and the Lovaas Institute anchored the provision of therapy based in applied behavioral analysis designed to "recover" autistic children, alongside the parent-based advocacy organization Cure Autism Now (which the similar organization Autism Speaks absorbed that year). Meanwhile, groups such as Talking About Curing Autism (TACA) and Generation Rescue (both represented around this time by Jenny McCarthy, "Ph.D. in Google Research" [6]) spread potentially deadly disinformation including vaccine skepticism and chelation therapy. Amid this hostile climate, I decided to earn an actual Ph.D. (in educational psychology) at the heart of the medical model of autism research, the University of California Los Angeles, to learn the thinking and language of mainstream science to better critique them. Over this time I observed early autistic leaders (most prominently Jim Sinclair) become much less active, and their products become increasingly difficult to access, such as websites no longer hosted or archived. Leaders of autism's cure movement took down materials as well, but this more often happened to obscure their more outrageous products in response to autistic-led resistance, as they usually have the resources to continue displaying them (see Rosenblatt [7] for examples). The need to preserve and document the history of the autistic community and neurodiversity movement had become apparent. Like my career, the book merges science and advocacy, intended for both academia and the autism community.

Exeter, UK   Steven K. Kapp

References

1. Bascom, J. (2012). *Loud hands: Autistic people, speaking.* Washington, DC: Autistic Self Advocacy Network.
2. Braun, V., & Clarke, V. (2006). Using thematic analysis in psychology. *Qualitative Research in Psychology, 3*(2), 77–101.
3. Brown, L. X., Ashkenazy, E., & Onaiwu, M. G. (2017). *All the weight of our dreams: On living racialized autism.* Lincoln, NE: DragonBee Press.
4. Gillespie-Lynch, K., Kapp, S. K., Pickens, J., Brooks, P., & Schwartzman, B. (2017). Whose expertise is it? Evidence for autistic adults as critical autism experts. *Frontiers in Psychology, 8*, 438.
5. Hughes, J. M. (2016). *Increasing neurodiversity in disability and social justice advocacy groups.* Washington, DC: Autistic Self Advocacy Network.
6. McCarthy, J. (2007). *Louder than words: A mother's journey into healing autism.* New York: Dutton.
7. Rosenblatt, A. (2018). Autism, advocacy organizations, and past injustice. *Disability Studies Quarterly, 38*(4). Retrieved from http://dsq-sds.org/article/view/6222.
8. Russell, G., Kapp, S. K., Elliott, D., Elphick, C., Gwernan-Jones, R., & Owens, C. (2019). Mapping the autistic advantage from the experience of adults diagnosed with autism: A qualitative study. *Autism in Adulthood* (Advance online publication).
9. Sequenzia, A., & Grace, E. J. (2015). *Typed words, loud voices.* Fort Worth, TX: Autonomous Press.

10. Silberman, S. (2015). *NeuroTribes: The legacy of autism and the future of neurodiversity*. New York: Avery.
11. Walker, N. (2014, September 27). Neurodiversity: Some basic terms and definitions. *Neurocosmopolitanism*.
12. Waltz, M. (2013). *Autism: A social and medical history*. Basingstoke, UK: Palgrave Macmillan.

Acknowledgements

I wish to acknowledge the generous support of the Wellcome Trust, the grant of which (108676/Z/15/Z) funded my time producing the book as part of the project Exploring Diagnosis: Autism and Neurodiversity at the University of Exeter. Thank you also to Dr. Ginny Russell, Daisy Elliott, and Rhianna White of the Exploring Diagnosis project. Ginny's inclusion of me and a book as part of the project made this specific edited collection possible, and her vision, work with me on the application, supervision of the process, and contributions critically supported the product's scholarship. Daisy's administrative support helped to keep me organized and allowed me to focus on the content. Rhianna's assistance with formatting and proofreading came in timely fashion, helping me to deliver the main manuscript. Also at the University, Professor Brahm Norwich provided invaluable, steady mentorship on editing the book. Too many people have shaped my development and career to acknowledge when thinking beyond the production of the book, but within that domain I would like to thank Josh Pitt of Palgrave Macmillan for commissioning the book and Sophie Li of Springer Nature for assisting the publication process with a positive approach, patience, and flexibility.

The autistic community members and activists deserve recognition not only for their contributions to the neurodiversity movement, but also thanks for the time and effort in producing heartfelt, well-written chapters. Ari Ne'eman provided input that shaped the direction of the book, most directly through helping with selecting and recruiting authors. Ari and Dr. Scott Michael Robertson created a vehicle for my activism through co-founding the Autistic Self Advocacy Network, and I am indebted to Scott for introducing me to the disability rights and neurodiversity movements and providing patient, knowledgeable mentorship. Nick Chown produced a professional index that benefitted from his insider status as an autistic neurodiversity scholar. Ryan Smoluk created the cover art, "The Path (Planning Alternative Tomorrows with Hope)", which offers a brilliant image from a member of the autistic community that resonates with work of the neurodiversity movement and spirit of the book.

Contents

1   Introduction   1
    Steven K. Kapp

Part I   Gaining Community

2   Historicizing Jim Sinclair's "Don't Mourn for Us":
    A Cultural and Intellectual History of Neurodiversity's
    First Manifesto   23
    Sarah Pripas-Kapit

3   From Exclusion to Acceptance: Independent Living
    on the Autistic Spectrum   41
    Martijn Dekker

4   Autistic People Against Neuroleptic Abuse   51
    Dinah Murray

5	Autistics.Org and Finding Our Voices as an Activist Movement Laura A. Tisoncik	65
6	Losing Mel Baggs	77

Part II Getting Heard

7	Neurodiversity.Com: A Decade of Advocacy Kathleen Seidel	89
8	Autscape Karen Leneh Buckle	109
9	The Autistic Genocide Clock Meg Evans	123
10	Shifting the System: AASPIRE and the Loom of Science and Activism Dora M. Raymaker	133
11	Out of Searching Comes New Vibrance Sharon daVanport	147
12	Two Winding Parent Paths to Neurodiversity Advocacy Carol Greenburg and Shannon Des Roches Rosa	155
13	Lobbying Autism's Diagnostic Revision in the DSM-5 Steven K. Kapp and Ari Ne'eman	167
14	Torture in the Name of Treatment: The Mission to Stop the Shocks in the Age of Deinstitutionalization Shain M. Neumeier and Lydia X. Z. Brown	195

15	*Autonomy, the Critical Journal of Interdisciplinary Autism Studies* Larry Arnold	211
16	My Time with Autism Speaks John Elder Robison	221
17	Covering the Politics of Neurodiversity: And Myself Eric M. Garcia	233
18	"A Dream Deferred" No Longer: Backstory of the First Autism and Race Anthology Morénike Giwa Onaiwu	243

Part III Entering the Establishment?

19	Changing Paradigms: The Emergence of the Autism/Neurodiversity Manifesto Monique Craine	255
20	From Protest to Taskforce Dinah Murray	277
21	Critiques of the Neurodiversity Movement Ginny Russell	287
22	Conclusion Steven K. Kapp	305
Index		319

List of Figures

Fig. 4.1	The chart clearly showed that the vast majority of claimed "benefits" of medication in this sphere are about reducing behavior rather than enhancing personal well-being or capacity	53
Fig. 4.2	Summarized comparisons of restrictive or potentially fatal effects of psychotropic drugs prescribed for autistic adults with learning disabilities and challenging behavior (ca 2001). From left to right, starting with the antipsychotics: phenothiazines (e.g. Chlorpromazine), thioxanthines (e.g. Fluenthixol), Haloperidol, pimozide, risperidone, olanzapine; and then various non-neuroleptic experimental drugs for autism-related perceived problems, viz paroxetine, lithium, carbamazepine, buspirone, naltrexone, and the beta-blocker propranolol	56
Fig. 4.3	The APANA logo, designed by Ralph Smith	58
Fig. 9.1	The Autistic Genocide Clock by Meg Evans	124
Fig. 11.1	Autistic Women and Non-Binary Network tagline	153
Fig. 20.1	Anti-Autism Speaks Logo designed by Dinah Murray	279
Fig. 20.2	Another Anti-Autism Speaks Logo circulating in 2008, anonymous	279

1

Introduction

Steven K. Kapp

This book marks the first historical overview of the autistic community and the neurodiversity movement that describes the activities and rationale of key leaders in their own words. All authors of the core chapters consider themselves part of the autistic community or the neurodiversity movement (including a couple among the growing legion of non-autistic parents), or both in most cases. Their first-hand accounts provide coverage from the radical beginnings of autistic culture to the present cross-disability socio-political impacts. These have shifted the landscape toward viewing autism in social terms of human rights and identity to accept, rather than as a medical collection of deficits and symptoms to cure. The exception to personal accounts and part of the impetus for the book, Jim Sinclair, has become inactive since leading the autism rights movement's development

---

S. K. Kapp (✉)
College of Social Sciences and International Studies, University of Exeter, Exeter, UK
e-mail: steven.kapp@port.ac.uk

Department of Psychology, University of Portsmouth, Portsmouth, UK

© The Author(s) 2020
S. K. Kapp (ed.), *Autistic Community and the Neurodiversity Movement*,
https://doi.org/10.1007/978-981-13-8437-0_1

of culture and identity after co-founding its first organization Autism Network International (ANI) in 1992. Yet this book respects the disability rights motto of "Nothing About Us Without Us" by commissioning an autistic historian and chairperson of an organization inspired by ANI's historic autistic community retreat to analyze the context and impact of Sinclair's legendary work. Similarly, I am an autistic neurodiversity activist (a role that precedes my career as an autism researcher), but I endeavor to apply robust scholarly standards to editing this collection (see the Preface).

Introduction to the Neurodiversity Movement

Many descriptions arguably misunderstand the concept of *neurodiversity* and the framework and actions of the *neurodiversity movement*, so this chapter seeks to explain them before introducing the core chapters.

The term *neurodiversity* originates from the autism rights movement in 1998 from Judy Singer on Martijn Dekker's mailing list InLv, but as the movement has matured into a more active part of a cross-disability rights coalition, the term has evolved to become more politicized and radical (a change noted by a few contributors, especially Dekker in Chapter 3). *Neurodiversity* has come to mean "variation in neurocognitive functioning" (p. 3) [1], a broad concept that includes everyone: both *neurodivergent* people (those with a condition that renders their neurocognitive functioning significantly different from a "normal" range) and *neurotypical* people (those within that socially acceptable range). The *neurodiversity movement* advocates for the rights of neurodivergent people, applying a framework or approach that values the full spectra of differences and rights such as inclusion and autonomy. The movement arguably adopts a spectrum or dimensional concept to neurodiversity, in which people's neurocognitive differences largely have no natural boundaries. While the extension from this concept to group-based identity politics that distinguish between the neurodivergent and neurotypical may at first seem contradictory, the neurodiversity framework draws from reactions to *existing* stigma- and mistreatment-inducing medical categories *imposed* on people that they

Reclaiming words imposed

reclaim by negotiating their meaning into an affirmative construct. People who are not discriminated against on the basis of their perceived or actual neurodivergences arguably benefit from neurotypical privilege [2], so they do not need corresponding legal protections and access to services. I have observed little serious aggrandizement of neurodivergent people or denigration of neurotypical people, but satire has been misinterpreted (Tisoncik, Chapter 5) or rhetoric misunderstood due to disability-related communication or class differences.

The Diversity in *Neurodiversity*

Although the people for whom the neurodiversity movement advocates far exceed autistic people, they also fall outside the main scope of the book. Some contributors' topics do include campaigns directly affecting people with other disabilities, such as that to close the tortuous Judge Rotenberg Center in the U.S. (Neumeier and Brown, Chapter 14) and to pass the Autism/Neurodiversity Manifesto in the U.K. (Craine, Chapter 19), yet the movement remains led by autistic people. Mainly though, the scope of the movement remains unclear; at a disability studies conference I asked participants how they felt about minimum criteria for eligibility within it, but they felt uncomfortable posing limits [3]. A woman suggested her multiple sclerosis should qualify; indeed, coverage of people with not only chronic illnesses but also primary sensory disabilities like blindness and psychiatric conditions like schizophrenia remain unclear [4]. One issue may be the importance of the cure issue to the movement; for example, an autistic neurodiversity activist advocates for acceptance for autism but a cure for epilepsy (which she sees as separate from her sense of self and understands as potentially fatal). Such neurological conditions fall within the broader disability rights movement and deserve basic rights accommodated, such as, arguably, policy to ban flash photography in public places that could trigger seizures in people with photosensitive epilepsy [5]. The primacy of biology to the movement seems clear due to the *neuro-* in *neurodiversity*, and debates as to whether relevant neurodivergences must be neurodevelopmental or can be acquired environmentally or in adulthood have taken place in the U.K. [6]. Conditions such as schizophrenia fall within another identity-based socio-political movement (the mad pride movement, and while the neurodiversity movement may help provide a bridge to the disability rights movement, many adherents do not view themselves as disabled [7]. More importantly and practically, campaigns to attribute these conditions to the brain have backfired, likely because the public often associates them with violence and thinks brain-based conditions are more difficult to treat [8, 9]. Ultimately, book contributors did not exclude any particular conditions from the domain of the movement, and the right to self-determination offers the opportunity for other people to identify and organize within the movement.

While some activists say *neurodiversity* refers simply to a biological fact of this variance as opposed to the movement [10, 11], contributors to this volume—as aware autistics do generally: see Preface—suggest the term implicitly refers to a tenet of inclusion based on universal rights principles, with an emphasis on those with neurological disabilities. This includes aspirations of full inclusion in education, employment, and housing; freedom from abuse (e.g. abolition of seclusion and both chemical—that is, overmedication to control behavior—and physical restraint); and the right to make one's own decisions with support as needed. Contributors evoke "the compassionate, inclusive flavor of the word" (Seidel, Chapter 7) and "human rights concept" (Greenberg, Chapter 12): "the specific premise of neurodiversity is full and equal inclusion…Neurodiversity is for everyone" (daVanport, Chapter 11). Buckle (Chapter 8) clarifies that this inclusion involves interaction between diverse groups even in settings prioritized around the needs of a particular group: neurodiversity "means having NTs [neurotypicals] in autistic space as much as it does autistics in NT space". Raymaker (Chapter 10) explains both parts of the compound word: "Neurodiversity, to me, means both a fabulous celebration of all kinds of individual minds, and a serious, holistic acknowledgement of the necessity of diversity in order for society to survive, thrive, and innovate", which as Garcia (Chapter 17) states requires that society "welcome neurodivergent people and give them the tools necessary to live a life of dignity". Inspired by the principles of other social justice movements, the neuro*diversity* movement recognizes intersectionality (how neurodivergent people's disadvantages are compounded by other types of social oppression) beyond cross-disability solidarity, such as race (see Giwa Onaiwu, Chapter 18), gender including gender identity (see daVanport, Chapter 11), and class (such as the call by Woods [2017] for universal basic income).

Like the far-reaching concept of *diversity*, the neurodiversity movement as applied to autism functions inclusively, in that activists include non-autistic people as allies, and it accepts and fights for the full developmental spectrum of autistic people (including those with intellectual disability and no or minimal language). Marginalization of non-autistic people by non-autistic relative-led autism organizations catalyzed the movement (Pripas-Kapit, Chapter 2; [12]). Thus it seeks to help families with advocacy for acceptance, understanding, and support that can positively impact people across the autism spectrum and their parents [13]. Celebratory acts for parents toward autistic children such as learning to speak their child's language and even accepting autism as part of their child's identity, and ameliorative acts like parents teaching their child adaptive skills to cope in wider society, both show nearly universal support among the autism community—including "pro-cure" parents and "pro-acceptance" autistic people [14], yet many of the more powerful parental organizations have behaved in dehumanizing and polarizing ways toward autistic people, such as using fear and pity as fundraising strategies and seeking an end to all autistic people regardless of their preferences (daVanport, Chapter 11). They have appropriated self-advocacy by using language such as "families with autism" (whereas if anyone "has" autism, autistic people do). They have also claimed autistic people cannot advocate for public policy affecting their children (even though some autistic activists themselves have intellectual disability, language impairment or no speech, epilepsy, gastrointestinal disorders, self-care needs such as toileting or daily living, meltdowns, etc., or their children do: [15, 16].

The *Neuro-* in *Neurodiversity*

While the neurodiversity movement generally views autism as natural and essentially innate, despite the inability of clinicians to identify it from birth, this viewpoint transcends politics despite its utility in activism. Autistic people tend to view autism as arising entirely from biological causes, with no evident influence from the movement [14]. This may occur both because autistic people likely cannot remember their life before autism becomes diagnosable, and because autistic people more often conceive of and describe autism from the inside, referring to internal processes such as thoughts, emotions, and sensations rather than behavior [17]. This conception of autism privileges lived experience, and complements autistic activists' arguments that underlying differences and difficulties persist despite coping mechanisms that may behaviorally "mask" autism, which have support from neuroscientific and other research [18]. Such a phenomenon helps autistic people counter the attack "You're not like my child" from parents; see the group blog We Are Like Your Child (http://wearelikeyourchild.blogspot.com/). It also facilitates a neurological kinship of sorts with fellow autistic people, helping us to emphasize within-group commonalities to develop a sense of community despite variability in how our behaviors present, and to argue for our rights based on what Silverman [19] calls "biological citizenship". An inside-out viewpoint of autism also helps advocates of neurodiversity explain adaptive reasons *why* autistic people engage in atypical behaviors, such as "stimming" (e.g. body rocking and hand flapping: Kapp et al. [20]; Schaber [21].

Importantly, brain-based explanations facilitate the movement's compatibility with alliances with non-autistic parents. They reject a role in caregiving for causing autism, absolving parents of the responsibility scientists and clinicians assign(ed) to them when Freudian psychogenic theories have dominated (as they still do in France and to a lesser extent in countries such as Brazil). This may reduce parents' aversion toward listening to neurodiversity advocates describe helpful parenting practices. Many of the more successful "therapeutic" approaches involve educating others to respectfully understand autistic people's differences, such as teaching *responsive* caregiving tactics to parents that require them to "learn to speak their child's language" and communicate on their terms [13]. Researchers developed these techniques based on successful positive parenting practices in general [22]. A model that allows more for environmental contributions to autism's causation might look like parent-blaming, sparking resistance, and stifling progress. Moreover, biological explanations argue against environmental toxins as a risk factor for autism, helping to direct parents away from cottage industries based on rejected and unproven theories that offer dangerous "treatments" like heavy metal-injecting chelation therapy, chemical castration (Lupron therapy) bleach enemas, and vaccine avoidance (amid other expensive or at least ill-conceived "interventions"). Instead, biological explanations led by the neurodiversity movement help to raise ethical concerns about the basic scientific research that dominate autism research (such as the possibility of eugenics; see Evans, Chapter 9).

Interaction with the Medical Model

Although many claim that the neurodiversity movement simply supports the social model of disability and opposes the medical model, neurodiversity activists instead acknowledge the transaction between inherent weaknesses and the social environment [23, 24]. The social model of disability distinguishes between the core *impairments* inherent to medicalized conditions and *disability* caused by societal barriers (e.g. lack of assistive technology and physical infrastructure to enable someone with a mobility disability to move where they want to go), which for autism especially include social norms that result in misunderstandings and mistreatment [25]. One of the social model originators Mike Oliver [26] explained that he never advocated it as all-encompassing or intended it to *replace* the individual (medical) model, but to serve as an academic-political tool to help empower disabled people by emphasizing attention to the social obstacles that unite us; that it has certainly done. Yet the impairment that the model separates from disability may certainly add to any individuals' struggles. In practice this means that the neurodiversity movement begins with its goal of quality of life, which includes but surpasses adaptive functioning (e.g. self-determination and rights, well-being, social relationships and inclusion, and personal development: Robertson [27]; see also Tsatsanis et al. [28]), and works backward from there to address the individual *and* social factors that *interact* to produce disability. In contrast, a "pure" medical model approach would assume an individual's "symptoms" (behaviors or traits) directly and specifically cause dysfunction or disability, and work to disrupt this linear relationship by preventing or curing the condition. Yet the disability rights movement has already helped enshrine access (e.g. reasonable accommodations) and non-discrimination into law, and medical practices have gradually changed to allow more patient and client autonomy [29]. Indeed, social and medical models have moved toward one another over time [24].

The neurodiversity movement's opposition to "curing" autism has produced misunderstandings, such as mistaken assumptions that it attributes all challenges to social injustices and rejects interventions to mitigate them. While the movement disagrees with certain principles, means, and goals of interventions, with those caveats, it does support therapies to help build useful skills such as language and flexibility. It opposes framing these matters in unnecessarily medical or clinical ways; arguably all interventions that have a scientific evidence base for truly helping autistic people's core functioning involve active learning (by the autistic person or others), and therefore one might describe them as "educational". It recognizes that some behaviors associated with neurodivergences like autism can serve as strengths (such as interests), as coping mechanisms for underlying differences that can prove challenging at times (such as forms of stimming like hand flapping and body rocking, which help to self-regulate and communicate overpowering emotions, among other functions: Kapp et al. [20]; Bascom [30]), or as inherently neutral differences (such as an apparently monotone voice or a preference for solitude: Winter [31].

While all social movements have more radical left wings, arguably the organized, politically mobilized autism rights branch of the neurodiversity movement largely practices critical yet reformist pragmatism rather than revolution. The movement in some ways supports a Western biomedical model more than autism's medical establishment and certainly more than autism's organized cure movement. For example, the neurodiversity movement's framework conceptualizes autism itself as purely biological, as opposed to resulting from dynamic genetic-environmental interplay (as the mainstream autism field believes and as most research suggests) or at least in part from toxins in the physical environment (as many "pro-cure" parents and their advocacy organizations have believed). Neurodiversity activists support traditional medicine for preventing and treating ill health, such as vaccines to prevent infectious diseases and (with the individual's consent) psychotropic medication to treat anxiety and depression (see Murray, Chapter 4), whereas beliefs in the likes of false and discredited vaccine-autism links have energized radical pro-cure activists, pseudoscience, and fringe medicine.

Neurodiversity supporters cling essentially to autism's diagnostic criteria when challenging even mainstream critics, as we support acceptance of official autism domains of atypical communication, intense and "special" interests, a need for familiarity or predictability, and atypical sensory processing, yet distinguish between those core traits and co-occurring conditions we would be happy to cure such as anxiety, gastrointestinal disorders, sleep disorders, and epilepsy. We, as do all of the authors for this book and the latest revisions of autism's official diagnoses ([32]; https://icd.who.int/), generally support a unified conception of the autism spectrum. Understanding and production of structural language now fall outside of autism's criteria (as a separate communication diagnosis), and neurodiversity activists have likewise supported efforts to expand access to language and communication but do not regard this as making someone "less autistic", unlike arguably most autism advocates. Autistic neurodiversity activists have defined critical autism studies not in terms of being critical of autism's existence (unlike many non-autistic thinkers outside the movement), but of the power dynamics that marginalize autistic scholars, pathologize autism, and overlook social factors that contribute to disability in autistic people [33]. While we support moving to an alternative identification system that recognizes autism's nuances ([34]; Kapp and Ne'eman, Chapter 13), such as strengths that can aide or add difficulties to autistic people's lives depending on myriad factors [35], the often fractious autism community united around the need to protect autistic people's access to diagnosis because of the practical services and supports medical classification can provide. While the psychiatric and clinical establishment sharply criticized the American Psychiatric Association's Diagnostic and Statistical Manual of Mental Disorders (DSM) for *adding* and *expanding* most diagnoses (increasing medicalization of everyday problems) in its latest revision (DSM-5) or for lacking validity [36], the neurodiversity movement's leading organization the Autistic Self Advocacy Network (ASAN) worked more closely with the DSM-5 than any other in the autism community to protect access to diagnosis (Kapp and Ne'eman, Chapter 13).

Self-Advocacy

The neurodiversity movement's approach holds autistic and neurodivergent people responsible not for the origin of our problems (social barriers exacerbating biological challenges), but for leading the effort to solve them. This position—responsibility for the "offset" but not "onset" of problems—aligns with the compensatory model of helping and coping according to an analysis [37] of a classic theoretical paper [38]. Other identity-based social justice movements such as the civil rights movement share this approach, which Brickman and colleagues viewed as arguably superior because it encourages people to seek help (because it does not blame people for problems), yet actively exert control over their lives. Yet while they say on page 372 that the model "allows help recipients to command the maximum possible respect from the social environment" and enables mobilization, people oriented this way put enormous pressure on themselves to solve problems they did not create, risking distressing strain. Indeed, campaigners in this book noted the financial and sometimes emotional sacrifices made for their activism (Murray, Chapter 4; Seidel, Chapter 7; see also Pripas-Kapit, Chapter 2). Movement activists do not think neurodivergences like autism excuse abusive behavior, and call it out (especially in intersectional ways to protect more disempowered community members), such as educating autistic men about consent in sexual relationships (Garcia, Chapter 17; [39]). In contrast, the medical model holds people responsible for *neither* the causation of *nor* the solution(s) to their problems, making them dependent [38], albeit medical and clinical clients in general have become increasingly empowered in practice [29].

History and Introduction to Contributors

I commissioned contributors who have made significant achievements to the development or maturation of the autistic community or the neurodiversity movement. I posed the same questions to all contributors for them to consider: why and how they got involved, how they carried out their contribution, whether it has accomplished what they intended, etc.

Contributors took different approaches to addressing these questions, and while I suggested a topic (originally limited to a particular action) and length, they negotiated their preferences and needs with me. I chose to prioritize content rather than style in my editing, giving substantive feedback on drafts but deemphasizing grammar and structure, especially considering contributors' wide-ranging educational and cultural backgrounds as well as communication abilities, to preserve the voices of the activists (see also Giwa Onaiwu, Chapter 18).

The chapters follow a chronological order that reveals patterns in the growth of the neurodiversity movement over time, a historical orientation that emphasizes where the movement and autism field have been most active: the U.S. and U.K. (the home of all contributors except Dekker, who lives in the Netherlands but also spends significant time in the U.K.). These countries have had exceptional roles in pioneering mother-blaming psychoanalytic child psychology that have unjustly blamed parents and sometimes removed autistic children from them, giving rise to the first autism advocacy organizations [40, 41]. Those parent-led organizations empowered both world-leading scientific research and pseudoscience to establish autism as a treatable developmental disability [42]. Yet these nations also arguably hosted the birth of the disability rights movement (in the U.S.), the social model of disability (the U.K.), and disability studies (arguably both countries; see Waltz, 2013). Hence autistic adults had more to resist and resources at their service in these contexts, with similarities in various other anglophone countries and nations with high English fluency. Furthermore, most activities of the neurodiversity movement have taken place online, where people can participate internationally. This organizational approach to the book not only reflects not wishing to oversimplify other national and cultural contexts (e.g. Germany or Israel) with single chapters, but also the limitations of where I have lived and my social networks.

Part I: Gaining Community

At a time when non-autistic parents dominated autism advocacy in the early 1990s, Sinclair (Chapter 2) led the launch of the movement and delivered its pro-acceptance manifesto mainly intended for parents, "Don't Mourn for Us", helping autistic people gain an identity and communicate in cyberspace (ANI-L) and in person (Autreat). In 1996, Martijn Dekker's e-mail list InLv provided an inclusive, autistic-hosted space that helped spawn new ideas such as the term *neurodiversity* (Chapter 3). By 1998 Autistic activists demonstrated their ability to partner and ally with parents and non-autistic professionals on early campaigns they led, such as Dinah Murray's "Autistic People Against Neuroleptic Abuse" (Chapter 4). Laura Tisoncik's autistics.org website launched that year and gave voice to injustices such as through satire like the Institute for the Study of the Neurologically Typical (Chapter 5), yet now "neurotypical" has become a common descriptor for people without neurological disabilities in medical studies. Protest campaigns in response to specific events and initiatives have mounted, such as Mel Bagg's Getting the Truth Out website created in 2005 in response to the Autism Society of America's fear-mongering Getting the Word Out (Chapter 6), along with ongoing efforts like Autistics Speaking Day in response to Communication Shutdown and Autism Acceptance Day and Month in response to their Autism Awareness counterparts. The movement has grown to create annual events by autistic activists not in specific response to those by non-autistic people, including Autistic Pride Day launched by Amy and Gareth Nelson in 2005 and the Disability Community Day of Mourning, begun by Zoe Gross in 2012 to remember those people with disabilities murdered by family members and try to prevent future filicide.

Part II: Getting Heard

These activities have helped raise consciousness that the neurodiversity movement, while arising to counter the exclusion and pathologization autistic adults felt by organizations and conferences run mainly by non-autistic parents, serves to create a world where autistic and other disabled people are free to be themselves in a respectful and inclusive society. Indeed, Kathleen Seidel (Chapter 7) has hosted neurodiversity.com as a non-autistic parent, without significant protests that an autistic does not own the domain name (Chapter 7). The historic archives, posts by autistic and non-autistic guests on debates or issues, and Seidel's counters to disinformation like false, dangerous treatments for and beliefs of causes of autism have demonstrated the movement's alliance with like-minded parents and impactful commitment to science.

Inspired by Sinclair's Autreat, Autscape (Buckle, Chapter 8) provides the longest-running ongoing example of physical "autistic space": an annual conference mostly by and for autistic people, which has demonstrated the possibilities and limits of inclusion. Beginning at a similar time, the Autistic Genocide Clock webpage publicized autistic people's fears of eugenics to prevent autism through the development of a genetic test for selective abortion, and its creator Meg Evans (Chapter 9) took it down early mainly because of the progress of the neurodiversity movement in changing attitudes toward acceptance. During the time span between the autism genocide clock being created (2005) and taken down (2011), ASAN led the movement's maturation from a sociocultural to a sociopolitical movement actively part of the disability rights coalition, organizing a protest against a cross-disability campaign that united autistic people with parents of autistic individuals and disability rights activists alike [43].

The Academic Autism Spectrum Partnership in Research and Education (AASPIRE) project has demonstrated the expertise of even lay autistic people as the leading provider of participatory autism research (Raymaker, Chapter 10), illustrating the growing reach of the neurodiversity movement, as have other developments. The Autistic Women and Non-Binary Network (AWN) has provided powerful advocacy for intersectional feminism, as exemplified by its recent selection by the U.S. Library of Congress for preservation of its website, giving access to archives for current and future generations of advocates (daVanport, Chapter 11).

The Thinking Person's Guide to Autism provides a network of pro-neurodiversity and pro-science information hosted by autistic and non-autistic parents, providing the neurodiversity movement with an influential alliance that helps to reach the critical demographic of non-autistic parents (Greenburg and Rosa, Chapter 12). ASAN consulted on the revision of autism's diagnosis in the DSM-5, marking a historic collaboration that substantially affecting the core criteria and accompanying text to help maintain access to autism diagnoses and therefore needed supports (Kapp and Ne'eman, Chapter 13).

Shain Neumeier and Lydia Brown (Chapter 14) have taken leading roles in activism to stop the electric use of shocks as "treatment", raising the profile of the issue and providing strong legal and ethical arguments that have assisted progress toward banning the tortuous practice.

Larry Arnold (Chapter 15) edits *Autonomy, the Critical Journal of Interdisciplinary Autism Studies*, a journal that not only advances the cause of autistic people as editors and authors of new academic studies, but also preserves key texts of the neurodiversity movement.

John Elder Robison (Chapter 16) served as the only autistic advisor to Autism Speaks, the world's most powerful autism organization and the main enemy of the movement, and his resignation from his attempts to serve as a moderating influence contributed to reforms that have begun to soften its most contentious practices [44].

A journalist based in Washington, DC has found that a story in which he "outed" and explained himself as a member of the autistic community has led to opportunities to explain autism and disability politics in neurodiversity-affirming ways, a warm reception that demonstrates the growing public interest in autism rights and acceptance (Garcia, Chapter 17).

Morénike Giwa Onaiwu (Chapter 18) describes the principles of and her experience in editing the first anthology of autistic people of color, which in part through its publication by AWN further demonstrates the neurodiversity movement's intersectional autism advocacy [45], amid the broader autism community and media that often implicitly associate autism with whiteness [46].

Part III: Entering the Establishment?

At the present time in which autism acceptance continues to reach new heights, the neurodiversity movement has edged closer to the autism establishment, although the current status looks uncertain. A couple of current examples from the U.K. illustrate this point. In Chapter 19, Craine tells the story of how, following endorsement of the Autism/Neurodiversity

Manifesto by the Labour Party's finance minister, Neurodiversity Labour was launched in February 2019. This organization, led by people with neurodivergences such as autism, ADHD, dyslexia, and dyspraxia, fights discrimination against neurodivergent people within the Labour Party and society. In addition, the National Autistic Taskforce (Murray, Chapter 20) seeks to help implement the U.K.'s principled but hardly enforced legislation such as the Autism Act 2009, which has provisions for the needs of autistic adults. This autistic-led taskforce prioritizes minimally verbal autistic people with high support needs. It grew out of the National Autism Project, which provides access to government consultations and contacts that could help achieve its aims. If the broader autism community, public, and levers of power attain a critical mass of understanding and support for the neurodiversity framework and movement, autistic people will lead advocacy for control of our own affairs.

References

1. Hughes, J. M. F. (2016). *Nothing about us without us: Increasing neurodiversity in disability and social justice advocacy groups.* Retrieved from https://autisticadvocacy.org/wp-content/uploads/2016/06/whitepaper-Increasing-Neurodiversity-in-Disability-and-Social-Justice-Advocacy-Groups.pdf.
2. Harp, B. [bev]. (2009, August 7). *Check list of neurotypical privilege: New draft* (Blog post). Retrieved from http://aspergersquare8.blogspot.com/2009/08/checklist-of-neurotypical-privilege-new.html.
3. Kapp, S. (2011). *Neurodiversity and progress for intercultural equity.* Paper presented at the meeting of the Society for Disability Studies, San Jose, CA.
4. Liebowitz, C. (2016, March 4). Here's what neurodiversity is—and what it means for feminism. *Everyday Feminism.* Retrieved from https://everydayfeminism.com/2016/03/neurodiversity-101/.
5. Asasumasu, K. [Neurodivergent K]. (2012, January 12). *Why I am not going to Portland Lindy Exchange* (Blog post). Retrieved from https://timetolisten.blogspot.com/2012/01/why-i-am-not-going-to-portland-lindy.html.
6. Arnold, L. (2017). A brief history of "neurodiversity" as a concept and perhaps a movement. *Autonomy, the Critical Journal of Interdisciplinary Autism Studies, 1*(5). Retrieved from http://www.larry-arnold.net/Autonomy/index.php/autonomy/article/view/AR23/html.

7. Graby, S. (2015). Neurodiversity: Bridging the gap between the disabled people's movement and the mental health system survivors' movement. In H. Spandler, J. Anderson, & B. Sapey (Eds.), *Madness, distress and the politics of disablement* (pp. 231–244). Bristol, UK: Policy Press.
8. Feldman, D. B., & Crandall, C. S. (2007). Dimensions of mental illness stigma: What about mental illness causes social rejection? *Journal of Social and Clinical Psychology, 26*(2), 137–154.
9. Read, J. (2007). Why promoting biological ideology increases prejudice against people labelled "schizophrenic". *Australian Psychologist, 42*(2), 118–128.
10. Silberman, S. (2015). *NeuroTribes: The legacy of autism and the future of neurodiversity*. New York: Avery.
11. Walker, N. (2014, September 27). *Neurodiversity: Some basic terms and definitions* (Blog post). Retrieved from http://neurocosmopolitanism.com/neurodiversity-some-basic-terms-definitions/.
12. Sinclair, J. (2012). Autism network international: The development of a community and its culture. In J. Bascom (Ed.), *Loud hands: Autistic people, speaking* (pp. 17–48). Washington, DC: The Autistic Press. Retrieved from http://www.autreat.com/History_of_ANI.html.
13. Kapp, S. K. (2018). Social support, well-being, and quality of life among individuals on the autism spectrum. *Pediatrics, 141*(Supplement 4), S362–S368.
14. Kapp, S. K., Gillespie-Lynch, K., Sherman, L. E., & Hutman, T. (2013). Deficit, difference, or both? Autism and Neurodiversity. *Developmental Psychology, 49*(1), 59–71.
15. Ballou, E. P. [chavisory]. (2018, February 6). *What the neurodiversity movement does—and doesn't—offer* (Blog post). Retrieved from http://www.thinkingautismguide.com/2018/02/what-neurodiversity-movement-doesand.html.
16. Rosa, S. D. R. (2019, January 27). *I'm the parent of a "severe" autistic teen* (Blog post). I oppose the National Council on Severe Autism. Retrieved from http://www.thinkingautismguide.com/2019/01/im-parent-of-severe-autistic-teen-i.html.
17. Gillespie-Lynch, K., Kapp, S. K., Pickens, J., Brooks, P., & Schwartzman, B. (2017). Whose expertise is it? Evidence for autistic adults as critical autism experts. *Frontiers in Psychology, 8*, 438.
18. Kapp, S. (2013, January 23). *ASAN statement on fein study on autism and "recovery"*. Autistic Self Advocacy Network. Retrieved from https://autisticadvocacy.org/2013/01/asan-statement-on-fein-study-on-autism-and-recovery/.

19. Silverman, C. (2008). Brains, pedigrees, and promises: Lessons from the politics of autism genetics. *Biosocialities, Genetics and the Social Sciences* (pp. 48–65). New York: Routledge.
20. Kapp, S. K., Steward, R., Crane, L., Elliott, D., Elphick, C., Pellicano, E., & et al. (2019). 'People should be allowed to do what they like': Autistic adults' views and experiences of stimming. *Autism* (Advance online publication). https://doi.org/10.1177/1362361319829628.
21. Schaber, A. [gemythest]. (2014, January 25). *Ask an Autistic #1—What is stimming?* (Video file). Retrieved from https://www.youtube.com/watch?v=WexCWZPJE6A&t=322s.
22. Schreibman, L., Dawson, G., Stahmer, A. C., Landa, R., Rogers, S. J., McGee, G. G., & et al. (2015). Naturalistic developmental behavioral interventions: Empirically validated treatments for autism spectrum disorder. *Journal of Autism and Developmental Disorders, 45*(8), 2411–2428.
23. den Houting, J. (2019). Neurodiversity: An insider's perspective. *Autism, 23*(2), 271–273.
24. Kapp, S. (2013). Interactions between theoretical models and practical stakeholders: the basis for an integrative, collaborative approach to disabilities. In E. Ashkenazy & M. Latimer (Eds.), *Empowering leadership: A systems change guide for Autistic college students and those with other disabilities* (pp. 104–113). Washington: Autistic Self Advocacy Network (ASAN). Retrieved from http://autisticadvocacy.org/wp-content/uploads/2013/08/Empowering-Leadership.pdf.
25. Woods, R. (2017). Exploring how the social model of disability can be re-invigorated for autism: In response to Jonathan Levitt. *Disability & Society, 32*(7), 1090–1095.
26. Oliver, M. (2013). The social model of disability: Thirty years on. *Disability & Society, 28*(7), 1024–1026.
27. Robertson, S. M. (2010). Neurodiversity, quality of life, and autistic adults: Shifting research and professional focuses onto real-life challenges. *Disability Studies Quarterly, 30*(1). Retrieved from http://www.dsq-sds.org/article/view/1069/1234.
28. Tsatsanis, K. D., Saulnier, C., Sparrow, S. S., & Cicchetti, D. V. (2011). The role of adaptive behavior in evidence-based practices for ASD: Translating intervention into functional success. *Evidence-based practices and treatments for children with autism* (pp. 297–308). New York: Springer, US.
29. Baker, D. L. (2011). *The politics of neurodiversity: Why public policy matters.* Boulder, CO: Lynne Rienner Publishers.

30. Bascom, J. (2012). Quiet hands. In J. Bascom (Ed.), *Loud hands: Autistic people, speaking* (pp. 177–182). Washington, DC: The Autistic Press. Retrieved from https://juststimming.wordpress.com/2011/10/05/quiet-hands/.
31. Winter, P. [Stranger in Godzone]. (2011, November 14). *Behavioral therapy—'normalization' vs 'teaching of skills'* (Blog post). Retrieved from http://strangeringodzone.blogspot.com/2011/.
32. American Psychiatric Association. (2013). *Diagnostic and Statistical Manual of mental disorders (DSM-5®)*. Washington, DC: Author.
33. Woods, R., Milton, D., Arnold, L., & Graby, S. (2018). Redefining critical autism studies: A more inclusive interpretation. *Disability & Society, 33*(6), 974–979.
34. Chown, N. & Leatherland, J. (2018). An open letter to Professor David Mandell Editor-in-Chief, autism in response to the article "A new era in Autism". *Autonomy, the Critical Journal of Interdisciplinary Autism Studies, 1*(5). Retrieved from http://www.larry-arnold.net/Autonomy/index.php/autonomy/article/view/CO1/html.
35. Russell, G., Kapp, S. K., Elliott, D., Elphick, C., Gwernan-Jones, R., & Owens, C. (2019). Mapping the autistic advantage from the experience of adults diagnosed with autism: A qualitative study. *Autism in Adulthood* (Advance online publication). https://doi.org/10.1089/aut.2018.0035.
36. Kamens, S. R., Elkins, D. N., & Robbins, B. D. (2017). Open letter to the DSM-5. *Journal of Humanistic Psychology, 57*(6), 675–687.
37. Kapp, S. (June 2014). *Models of helping and coping with autism*. Paper presented at the meeting of the Society for Disability Studies, Minneapolis, MN.
38. Brickman, P., Rabinowitz, V. C., Karuza, J., Coates, D., Cohn, E., & Kidder, L. (1982). Models of helping and coping. *American Psychologist, 37*(4), 368–384.
39. Smith, J. [Joelle] (2012, August 15). *Please, don't be that guy!* (Blog post). Retrieved from https://evilautie.org/2012/08/15/please-dont-be-that-guy/.
40. Evans, B. (2017). *The metamorphosis of autism: A history of child development in Britain*. Manchester: Manchester University Press.
41. Silverman, C. (2011). *Understanding autism: Parents, doctors, and the history of a disorder*. Princeton, NJ: Princeton University Press.
42. Sweileh, W. M., Al-Jabi, S. W., Sawalha, A. F., & Sa'ed, H. Z. (2016). Bibliometric profile of the global scientific research on autism spectrum disorders. *SpringerPlus, 5*(1), 1480.
43. Kras, J. F. (2010). The "ransom notes" affair: When the neurodiversity movement came of age. *Disability Studies Quarterly, 30*(1). Retrieved from http://www.dsq-sds.org/article/view/1065/1254.

44. Rosenblatt, A. (2018). Autism, advocacy organizations, and past injustice. *Disability Studies Quarterly, 38*(4). Retrieved from http://dsq-sds.org/article/view/6222/5137.
45. Strand, L. R. (2017). Charting relations between intersectionality theory and the neurodiversity paradigm. *Disability Studies Quarterly, 37*(2). Retrieved from http://dsq-sds.org/article/view/5374/4647.
46. Heilker, P. (2012). Autism, rhetoric, and whiteness. *Disability Studies Quarterly, 32*(4). Retrieved from http://dsq-sds.org/article/view/1756/3181.

Open Access This chapter is licensed under the terms of the Creative Commons Attribution 4.0 International License (http://creativecommons.org/licenses/by/4.0/), which permits use, sharing, adaptation, distribution and reproduction in any medium or format, as long as you give appropriate credit to the original author(s) and the source, provide a link to the Creative Commons license and indicate if changes were made.

The images or other third party material in this chapter are included in the chapter's Creative Commons license, unless indicated otherwise in a credit line to the material. If material is not included in the chapter's Creative Commons license and your intended use is not permitted by statutory regulation or exceeds the permitted use, you will need to obtain permission directly from the copyright holder.

Part I
Gaining Community

2

Historicizing Jim Sinclair's "Don't Mourn for Us": A Cultural and Intellectual History of Neurodiversity's First Manifesto

Sarah Pripas-Kapit

In reviewing the intellectual history of neurodiversity, Jim Sinclair's 1993 essay "Don't Mourn for Us" stands out as almost singularly influential [1]. The essay was first published in the third-ever issue of *Our Voice*, the newsletter of Autism Network International (ANI). Sinclair, an ANI co-founder, based the essay on a presentation xe delivered at the 1993 International Conference on Autism in Toronto. The essay implored parents not to mourn for their autistic child's disability, but rather to embrace their child's differences and work to meet their needs.

Even nearly thirty years after its original publication, "Don't Mourn for Us" remains a touchstone for the neurodiversity movement, cited in both casual conversations on social media as well as more academic pieces offering cultural commentary and criticism.

The essay has served as a springboard for conversations about parental expectations in the context of an autism diagnosis. Many autistic people and parents cite the piece as leading them toward a path of self-acceptance

S. Pripas-Kapit (✉)
Bellevue, WA, USA

© The Author(s) 2020
S. K. Kapp (ed.), *Autistic Community and the Neurodiversity Movement*,
https://doi.org/10.1007/978-981-13-8437-0_2

[2, 3]. A few activists have critiqued Sinclair for not going far enough [4]. Conversely, some parents have criticized Sinclair for an alleged failure to understand their perspective [5, 6].

As a historian, I am less interested in arguing about the correctness of Sinclair's views here—although as an autistic person and advocate for neurodiversity, I agree with them. Rather, I'd like to illuminate the historical context of "Don't Mourn for Us." This piece will explore how Sinclair's work fits into the broader history of autistic people's advocacy and public speech.

In the interest of full disclosure, I don't just come at this from the perspective of a historian. I attended the event Sinclair founded, Autreat (ANI) in 2008 and 2010. While in attendance, I briefly met Sinclair. Currently, I am the chairperson of the Association for Autistic Community (AAC). We sponsor an autistic community retreat, Autspace, that continues many of the same traditions of Autreat (which met for the last time in 2013). These experiences have undoubtedly shaped my perspective on "Don't Mourn for Us" and Sinclair's place in the neurodiversity movement, though this piece is primarily intended as a historicization of Sinclair's body of work.

To historicize "Don't Mourn for Us," I will begin by looking at Sinclair's contemporaries. The mid-1980s and early 1990s saw some of the first published writings by autistic people in the English-speaking world, including the works of Temple Grandin and Donna Williams. By looking at Grandin's and Williams' writings, we can better understand the radicalism of "Don't Mourn for Us."

Within this context, I will then analyze Sinclair's intellectual evolution as seen through xyr public writings. Finally, I will suggest how Sinclair and "Don't Mourn for Us" have shaped the neurodiversity movement since 1993—and how the movement has developed since.

Autistic Writings and the Neurotypical Audience

The first autistic people to write for a wide English-speaking audience were Temple Grandin and Donna Williams. Grandin's autobiography,

Emergence: Labeled Autistic was first published in 1986 by Arena Press, and Williams' *Nobody, Nowhere*, was first published in Great Britain in 1991 and in the U.S. in 1992. *Nobody, Nowhere* became an international bestseller.

Both memoirs were radical in the sense that they introduced neurotypical audiences to the idea that autistic people could narrate their own experiences and had rich internal lives. Yet they were written for a neurotypical audience, and that shaped numerous aspects of the books' publication and content.

Grandin, who was born in the U.S. in 1947 and raised in an affluent white family, enjoyed several privileges that many autistic people of her generation lacked. She was not, however, diagnosed as autistic until she was a teenager. Her initial diagnosis was "brain damage," which likely saved her from being institutionalized as a child (Silberman [7]). However, Grandin was somewhat elusive on this point in the book's narrative. As the title suggests, *Emergence* presents the narrative that Grandin was able to "emerge" (or recover) from autism.

Although this is not made explicit in the book's text, Margaret M. Scariano was listed as a co-author of *Emergence: Labeled Autistic*. Scariano's role in shaping the book is unclear, although Grandin has since authored many published works as sole author. I have not been able to find much information about Scariano, although one 2013 obituary states that she wrote or co-wrote a number of books, including both fiction and non-fiction [8]. But while the circumstances behind Scariano's contributions remain ambiguous, the book's promotion of a "recovery" narrative is clear. In my analysis of *Emergence*, I have chosen to focus on this issue rather than the complexities of the book's authorship.

This emphasis on recovery is made explicit in the book's introduction, written by noted autism scientist Bernard Rimland (Rimland became famous for debunking the refrigerator theory of autism causation). In the introduction, Rimland recounts his acquaintance with Grandin, reassuring readers that she was "really" autistic—or, in his terms, "a recovered autistic individual" (Grandin & Scariano [9]). Rimland, who later became an advocate of dubious biomedical "treatments" for autism, gushed about Grandin's ongoing recovery in the introduction. The memoir's framing hence implicitly became something of a how-to guide for autism recovery.

In addition to Rimland's introduction, the book also includes a preface from William Carlock, who taught Grandin at a private school as an adolescent. These two introductory pieces served the function of "proving" Grandin's autistic status to a skeptical audience, while simultaneously suggesting that recovery from autism was both possible and desirable.

Grandin herself suggested this narrative in the text. *Emergence* is rife with descriptions of her autistic differences, including sensory sensitivities, communication differences, and other autistic traits. Grandin, a successful animal behavior scientist, used scientific terminology to explain autistic differences. But one cannot help but be left with the impression that autism is a tragedy. She described her alleged regression into autism at the age of six months:

> Mother, who was only nineteen when I was born, said she remembers me as a normal, healthy newborn with big blue eyes, a mass of downy brown hair, and a dimple in my chine. A quiet, 'good' baby girl named Temple.
>
> If I could remember those first days and weeks of life, would I have known I was on a fast slide slipping into an abyss of aloneness? Cut off by overreactions or inconsistent reactions from my five senses? Would I have sensed the alienation I would experience because of brain damage suffered as an unborn child—the brain damage that would become apparent in life when that part of the damaged brain matured? (Grandin 2005)

Significantly, Grandin did not actually remember any of the events recounted here, since they purportedly occurred when she was a mere six months old. Rather, she created this narrative using her mother's memories and a paradigm of autism in which autism entraps autistic people into a world of isolation and misery.

Although Grandin did not take a firm stance on the always-contentious issue of autism causation, she was quite unequivocal in suggesting that autism was a tragedy not just for her, but for her entire family. The very first line of Grandin's own text stated, "I remember the day I almost killed my mother and younger sister, Jean" (p. 21)[9].

Yet the actual incident Grandin referenced was decidedly more prosaic than this dramatic opening suggests. Grandin went on to describe how, as a child, she threw an uncomfortable hat out of an open window in her mother's car. This caused her mother to lose control of the vehicle—certainly dangerous, but hardly the dramatic tragedy first implied. From the very beginning of the text, the tone is set. Autism is dangerous, even when manifested in seemingly trivial things such as disliking an itchy hat.

Nobody, Nowhere is superficially quite different from *Emergence*. It reveals the perspective of a decidedly less privileged autistic woman. Williams, who was Australian, was not diagnosed with autism as a child. Like many other autistic people who grew up in the 1960s and 1970s, she received alternative diagnoses, including psychosis. Williams discovered that she was autistic as an adult, after going through many years of familial abuse, homelessness, and domestic abuse in relationships with men. Her lyrical prose gives the book a very different reading experience than *Emergence*.

Yet despite these important differences, there are several similarities between the works. *Nobody, Nowhere* also began with a forward from Bernard Rimland, which echoed many of the same themes as the *Emergence* introduction. Rimland praised the book for providing inside insights into autism, which he valued as a researcher and as a parent. He explained, "Much of what Donna Williams has written about the experience of autism was already familiar to me—at an intellectual level. But *Nobody, Nowhere* provides a heretofore unavailable—and alarming—highly subjective appreciation of what it's like to be autistic" [10]. According to such non-autistic "experts," autistic people's internal experiences were inherently alarming.

A second introduction, written by Australian psychologist Lawrence Bartak, also appeared in *Nobody, Nowhere*. Bartak discussed autism from a clinical perspective at length. As with *Emergence*, the multiple introductions essentially suggested that autistic people can't be fully trusted to narrate autism. Their words must first be contextualized by non-autistic "experts" who can attest to the narrative's authenticity.

This is not to say, however, that autism is presented identically in the two narratives. While *Emergence* suggested that autism trapped Grandin in an unpleasant world of isolation, Williams admitted that she enjoyed being "in her own world" at times.

Although many autistic people have since come to interrogate the notion of autistic people as being trapped in their own worlds, at the time of *Nobody, Nowhere*'s publication it remained a dominant paradigm. Williams utilizes the paradigm in many interesting ways, writing:

> Everything I did, from holding two fingers together to scrunching up my toes, had a meaning, usually to do with reassuring myself that I was in control and no one could reach me, wherever the hell I was. Sometimes it had to do with telling people how I felt, but it was so subtle it was often unnoticed or simply taken to be some new quirk that 'mad Donna' had thought up. [10]

Hence, Williams showed that her autistic chances—even ones that were thought of as "mad Donna"—served a meaningful purpose for her. In this way she anticipated many of the ideas of the neurodiversity movement, including the popular notion that "behavior is communication."

However, *Nobody, Nowhere* hardly rejected the autism-as-tragedy paradigm in its entirety. Williams explained how she found the world as so hostile as a child that she created two personas to help her, Carole and Willie. She explained:

> I had created an ego detached from the self, which was still trapped by crippled emotions. It became more than an act. It became my life, and as I had to reject all acknowledgment of an emotional self, I had to reject all acknowledgment of Donna. I eventually lost Donna and became trapped in a new way. [10]

Although Williams acknowledged that her response was in large part shaped by the abuse and rejection she suffered, the narrative as a whole suggested that entrapment—either in her one world or in a fictional persona of her own creation—was the inevitable result of autism. She frequently referred to herself and other autistic people as "trapped and frightened" [10]. Like Grandin in *Emergence*, Williams expressed the hope that her account would help others.

Emergence: Labeled Autistic and *Nobody, Nowhere* were not specifically written for a parent audience. Yet in an era where first-person accounts of autism were so scarce, it is highly likely that they played an important role in the then-small autism parent community—the intended audience for "Don't Mourn for Us."

Given the centrality of parents in "Don't Mourn for Us," it is worth examining how parents are discussed in Grandin and Williams' earlier works. For Grandin, her mother, referred to simply as "Mother" in the text, was a near saint-like figure. Excerpts from Eustacia Cutler Grandin's journal are presented at several points in the book, along with several letters to teachers and medical professionals. Even today, the two women frequently make public appearances together. Grandin has often credited her mother's decisions for her own success. (Her father appeared much less frequently in the text, reflecting the 1950s gender roles that shaped Grandin's upbringing.)

Today many autistic activists would criticize Eustacia Cutler's parenting methods, which included admonishing her young daughter for a failure to make eye contact. But Grandin herself never wavered in her admiration, presenting Cutler as a loving, no-nonsense mother who provided sage advice to Grandin throughout her life while also advocating for her needs. Indeed, many people who have heard Grandin speak in recent years—myself among them—have noted Grandin's tendency to present a nostalgic view of her childhood. This includes her mother's strict style of parenting, which is particularly jarring to twenty-first-century audiences.

In strong contrast to Grandin, Williams criticized her family in *Nobody, Nowhere*. She movingly described the abuse she suffered at the hands of her mother and brother: "To them, I was a retard, a nut, a spastic. I threw 'mentals' and couldn't act normal. 'Look at her, look at her,' they would say about a child who, to them, was either a 'retard' when I was in my own world or a 'nut' when I was in theirs. I couldn't win" [10].

However, Williams also empathized with her family. She went on to write, "Looking at it from their point of view, I guess they couldn't win, either. My brother had probably woken up to the fact that I hardly acknowledged, let alone accepted, him" [10].

In Williams' narrative, the abuse she suffered at the hands of her family becomes understandable, though not quite acceptable. Although she rejected the idea that her mother's coldness had caused her cognitive differences, rejecting the "refrigerator mother" hypothesis, she concluded that her disability likely impacted her mother for the worse.

Williams wrote, "Though [my mother] was probably a social cripple before I was born, I accept my share of the responsibility for making her one, and for robbing both her and my brother of a free, more independent relationship with each other" [10]. In a mere sentence, Williams hence reiterated an incredibly harmful view of her disability that she has internalized—the notion that her own differences caused discord within her family. According to this formulation, she was largely responsible for their problems, which included abuse from both parents and Williams' mother's alcoholism. This abusive home environment would eventually lead Williams to leave her home at the age of fifteen.

Emergence and *Nobody, Nowhere* might have challenged the pernicious view that autistic people lacked thoughts and feelings, but the narratives reinforce another idea: that having an autistic child is a tragedy for families.

Both Grandin and Williams offered suggestions for parents, both implicitly and explicitly. *Emergence*, for example, includes a final chapter with a bulleted list of suggestions for helping autistic children. By and large these were practical suggestions involving sensory sensitivity, diet, and related issues. Although a few tips gestured toward the direction of self-acceptance, neither writer suggested that autism was anything other than a disability to be mourned.

Then Jim Sinclair came along.

The Radicalism of Sinclair

"Don't Mourn for Us" came out of the autism culture of the 1980s and early 1990s. In fact, Sinclair was friends with Williams, whom xe met on an early online mailing list for the parents of autistic people. Sinclair, Williams, and Xenia Grant formally founded ANI in February of 1992, when Williams visited the U.S. from Australia to promote *Nobody, Nowhere* [11]. Initially, ANI began as a pen pal program and a newsletter—the same newsletter which published "Don't Mourn for Us" one year later.

It is clear from Sinclair's writings in the early 1990s and later that xe was immersed in the world of autism parents and professionals. Autreat, the retreat Sinclair ran for more than fifteen years, grew out of these experiences. Sinclair, Williams, and other autistic people who attended non-autistic-run conferences found them to be inaccessible, prohibitively expensive, and sometimes downright dehumanizing.

There were a few exceptions. In 1992, Sinclair wrote very positively of a TEACCH conference xe attended in 1989 [12]. It speaks to the paucity of such events at the time that Sinclair drove 1200 miles to attend the conference. The essay, entitled, "Bridging the Gaps: An Inside-Out View of Autism (Or, Do You Know What I Don't Know?)," was published in a TEACCH anthology that also included pieces from non-autistic experts Lorna Wing and Catherine Lord [7, 12].

This 1992 piece —cited much less frequently than "Don't Mourn for Us"—provides interesting glimpses into Sinclair's intellectual evolution. More so than any of Sinclair's subsequent writings, this essay included discussion of Sinclair's personal experiences and autism-related impairments. As the parenthetical part of the title suggests, Sinclair presented the autistic experience largely as an experience of not knowing. This not-knowing experience encompassed both not knowing the norms of the neurotypical world and not knowing about one's own autistic differences.

Sinclair reflected on xyr own experiences as an autistic child who grew up in the 1960s and 1970s. They explained:

> I've been living with autism for 27 years. But I'm just beginning to learn about what that means. I grew up hearing the word but never knowing what was behind it. My parents did not attend programs to learn about autism, did not collect literature to educate schools about autism, did not explain, to me or to anyone else, why my world was not the same one that normal people live in. [12]

For Sinclair, this feeling of isolation and not-knowing started to dissipate upon attending autism conferences. These conferences included a small number of other autistic adults. Yet at the same time, the autism conference world introduced Sinclair to a new type of isolation: being seen as an Other by non-autistic parents and professionals. To them, autistic people's experiences were something to be studied under a microscope, like an unusual virus. (Indeed, one sees evidence of this attitude in Rimland's introduction to Grandin's and Williams' narratives.) Sinclair did not care for such attitudes.

In the article, Sinclair took care to dispel myths about autistic people. The article's first subheading is "Being Autistic Does Not Mean Being Mentally Retarded," in a point that reads to many contemporary autistic activists as problematic in its failure to extend solidarity toward people with intellectual disabilities. Other subheadings are "Being Autistic Does Not Mean Being Uncaring" and "Being Autistic Will Always Mean Being Different."

But being different was not bad to Sinclair. "Bridging the Gaps" included several hints at the neurodiversity ideology that Sinclair would later articulate more fully. Xe started to articulate the idea that autistic people's impairments largely stemmed from societal factors, not inherent deficits.

Sinclair pointed to non-autistic people's assumptions as a key factor that limited autistic people. Xe criticized the special education field for being particularly unwilling to extend understanding toward autistics. Xe wrote:

> Not all the gaps are caused by my failure to share other people's unthinking assumptions. Other people's failure to question their assumptions creates at least as many barriers to understanding. The most damaging assumptions, the causes of the most painful misunderstandings, are the same now as they were when I was a child who couldn't talk, a teenager who couldn't drive, and a college student who couldn't get a job: assumptions that I understand what is expected of me, that I know how to do it, and that I fail to perform as expected out of deliberate spite or unconscious hostility.
>
> Other people's assumptions are usually much more resistant to learning than my ignorance. As a graduate student I encountered these assumptions in employers who had extensive backgrounds in special education. [12]

Sinclair did not specifically reference the social model of disability (which was much less well-known in 1992 than it is today, even in disability circles). However, xe did suggest a view of autism congruent with the social model.

These ideas would take a more fully realized view in "Don't Mourn for Us" one year later. In this piece, which has been referred to as a manifesto for the neurodiversity movement, Sinclair did not blunt xyr criticisms of parents. Xe focused on the parental tendency to "mourn" a child's autistic status. Sinclair stated,

> Parents often report that learning their child is autistic was the most traumatic thing that ever happened to them. Non-autistic people see autism as a great tragedy, and parents experience continuing disappointment and grief at all stages of the child's and family's life cycle.
>
> But this grief does not stem from the child's autism in itself. It is grief over the loss of the normal child the parents had hoped and expected to have. [1]

With this declaration, Sinclair identified the source of parental grief over having an autistic child. The fundamental cause was not the inherent tragedy of disability, but rather the pernicious cultural assumption that parents ought to have a "normal" child.

Such an assumption, Sinclair wrote, was damaging to both the parent and child. Although xe acknowledged that "Some amount of grief is natural as parents adjust to the fact that an event and a relationship they've been looking forward to isn't going to materialize," Sinclair urged parents to move beyond those feelings. Xe stated simply, "Continuing focus on the child's autism as a source of grief is damaging for both the parents and the child, and precludes the development of an accepting and authentic relationship between them" [1].

Such sentiments are radical even today. But when we consider the relevant historical context, they become even more so. Prior to Sinclair's declaration, the default mode for autistic people discussing autism was to focus almost exclusively on their personal experiences. Even if they admitted to enjoying parts of the autistic experience (as did Donna Williams), previous autistic writers always made sure to acknowledge the pain and danger that autistic people inflicted upon family members. Sinclair told parents that their feelings of grief, while very real, weren't the result of autism per se.

"Don't Mourn for Us" also dispelled the myth of the autistic person as being in their own world, another trope that appeared prominently in Grandin and Williams' work. Sinclair explained:

> You try to relate to your autistic child, and the child doesn't respond. He doesn't see you; you can't reach her; there's no getting through. That's the hardest thing to deal with, isn't it? The only thing is, it isn't true.
>
> Look at it again: You try to relate as parent to child, using your own understanding of normal children, your own feelings about parenthood, your own experiences and intuitions about relationships. And the child doesn't respond in any way you can recognize as being part of that system.
>
> That does not mean the child is incapable of relating at all. It only means you're assuming a shared system, a shared understanding of signals and meanings, that the child in fact does not share. [1]

In rejecting the "own world" paradigm, Sinclair also offered practical advice to struggling parents. But xe asked parents to understand their child's perspective rather than impose their own preferences and perspective on the child.

Although "Don't Mourn for Us" has oftentimes been interpreted as being dismissive of parental perspectives, Sinclair explicitly acknowledged the reality of parental grief, and the array of impairments that autistic people can experience. However, xe strenuously argued that parental grief should not be directed at the child. Xe wrote, "You didn't lose a child to autism. You lost a child because the child you waited for never came into existence. [...] Grieve if you must, for your own lost dreams. But don't mourn for us. We are alive. We are real. And we're here waiting for you" [1].

The essay also included commentary about how parent-run autism organizations could reorient themselves to better reflect autistic people's needs and priorities. Sinclair went on to suggest, "this is what I think autism societies should be about: not mourning for what never was, but exploration of what is. We need you. We need your help and your understanding. Your world is not very open to us, and we won't make it without your strong support" [1]. Xe hence invited parents to join autistic adults in creating a better world for autistic people—but not by demanding that autistic people "recover." Rather, Sinclair's vision of neurodiversity prioritized the reshaping of social expectations and norms. Non-autistic parents had a place in this movement, but it was primarily as allies to autistic adults (in the parlance of today's social justice vocabulary).

By choosing to take this radical stance, Sinclair sacrificed much. A friend of mine who conversed with xem on the subject said that Sinclair was on track to become a professional autistic speaker akin to Grandin and Stephen Shore. After taking more radical stances on autism and neurodiversity, those opportunities were no longer open. For a time, Sinclair was homeless. Xe never found a full-time job in xyr chosen profession as a rehabilitation counselor despite obvious knowledge and qualifications.

Yet Sinclair's sacrifices have borne considerable fruit. Although much has changed since the essay's original publication, the core idea articulated in "Don't Mourn for Us" has continued to animate neurodiversity activism. Sinclair changed the paradigm with which autistic adults would approach public speech. No longer were autistic people limited to personal narratives that relied heavily on tropes of autism as tragedy or entrapment. Autistic people could—and would—articulate their own views independent of parent and professional validation. That is the legacy of "Don't Mourn for Us."

Sinclair, Autspace, and the Development of Autistic Culture

Autistic culture as it exists today would be very different if not for the considerable contributions of Jim Sinclair. However, autistic culture and the philosophy of neurodiversity have undergone substantial shifts since 1993.

One of the most notable features of Sinclair's early work is the extent to which it began as a response to autism parent and professional culture. In some ways, this is seen even in Autreat, which Sinclair designed to be an autistic space that prioritized autistic needs [11].

Take, for example, Autreat's famous "Ask a Neurotypical" panel. According to Sinclair's description at one of the Autreats I attended, the idea for the panel originated as a parody of sorts. Sinclair disdained the "ask an autistic" panels frequently found at conferences for parents and participants. Xe had participated in many such panels, in which autistic panelists were asked entirely inappropriate questions such as "do you have sex?"

Xyr idea was to turn the tables. At the "Ask an NT" panel, autistic audience members would ask non-autistic panelists the same sorts of questions. This subversive idea, focused as it was on flipping the script, was fairly characteristic of the approach Sinclair took in "Don't Mourn for Us" and throughout xyr other works.

However, xe ran into a problem when trying to implement this plan. The autistic attendees at Autreat felt that the idea was unethical, premised as it was on asking people invasive questions in public without advance warning. So Sinclair scrapped the idea and the "Ask an NT" panel turned into something very different—an opportunity for autistic adults to learn more about non-autistic perspectives in a non-judgmental environment. (In one example of this dialogue, Sinclair asked panelists why neurotypicals enjoy eating at restaurants. Aside from the obvious pleasures of someone else cooking food for you, Sinclair asked, why bother with it?)

The evolution of the "Ask an NT" panel is in some ways emblematic of autistic culture's historical trajectory. Although it originated as a response to parents and professionals, it has since grown and mutated to develop its own traditions and community norms. Certainly Sinclair played a major role in the development of autistic culture, but always in dialogue with other autistic people.

Sinclair and ANI, the organization xe co-created, did not directly engage in policy advocacy. Yet the philosophy xe established would form the foundations of today's autistic-led policy advocacy work. Ari Ne'eman, co-founder of the Autistic Self-Advocacy Network (ASAN), explained it to me this way: "I never would've founded ASAN if not for Jim. ASAN might have popularized neurodiversity, but Jim Sinclair created it" (personal communication, February 20, 2019).

It is this fundamental idea that is the greatest legacy of "Don't Mourn for Us." Popular autism narratives of the 1980s and early 1990s suggested that autistic people were primarily useful for our ability to provide "inside insights" into the autistic experience. Sinclair transformed the paradigm by suggesting that autistic people could articulate a larger vision for social change. And we could do so without capitulating to the notion that autism was inherently tragic.

Sinclair's intellectual legacy extends well beyond "Don't Mourn for Us." Xe was likely the first autistic person to reject person-first language, in a 1999 essay "Why I Dislike Person First Language" [13]. Xe also coined the term "self-narrating zoo exhibit," which described the tendency of non-autistic parents and professionals to solicit personal narratives—like Grandin's and Williams'—that treated autistic people as peculiar curiosities. In all of xyr work, Sinclair was uncompromising in xyr willingness to question dominant narratives of autism as created by both non-autistic experts and less radical autistic representatives—the ones who were more likely to get conference invitations and book contracts.

Given Sinclair's emphasis on questioning all received wisdom, there is a certain irony to the now canonical status of "Don't Mourn for Us." The piece is certainly deserving of such status, but I believe Sinclair would be the first to admit that it was by no means intended as the final word on neurodiversity as a philosophy. It's particularly important to note that Sinclair's early work was shaped heavily by parent- and professional-dominated autism culture—a necessary move at the time xe first wrote the essay. Fortunately, we have now reached a point where it is possible to start creating more of our own cultural and intellectual traditions—a process which Sinclair began. Moving forward, I'd propose that future generations of autistics embrace the spirit of Sinclair's work by continuing to question, to challenge, and to move forward with new and innovative ideas.

References

1. Sinclair, J. (1993). Don't mourn for us. *Our Voice, 1*(3). Retrieved from http://www.autreat.com/dont_mourn.html.
2. Darroch, G. (2008, July 30), *Grief* (Web log post). Retrieved February 21, 2019 from http://autisticdad.blogspot.com/2008/07.
3. thequestioningaspie. (2016, October 31). *Revisiting Jim Sinclair—"Don't mourn for us": And (some of) my thoughts on the language of "cure"* (Web

log post). Retrieved February 21, 2019 from https://thequestioningaspie. wordpress.com/2016/10/31/revisiting-jim-sinclair-dont-mourn-for-us-and-some-of-my-thoughts-on-the-language-of-cure.
4. abfh. (2006, August 9). *Don't mourn, get attitude* (Web log post). Retrieved from http://autisticbfh.blogspot.com/2006/08/dont-mourn-get-attitude.html.
5. Morris, S. (2012, January 6). *An open letter to Jim Sinclair* (Web log post). Retrieved February 21, 2019 from http://sharon-theawfultruth.blogspot.com/2012/01/letter-to-jim-sinclair.html.
6. Thea ZebraMama. (2009, July 30). *My response to "Don't mourn for us" by Jim Sinclair* (Web log post). Retrieved February 21, 2019 from http://roosclues.blogspot.com/2009/07/my-response-to-dont-mourn-for-us-by-jim.html.
7. Silberman, S. (2015). *Neurotribes: The legacy of autism and the future of neurodiversity*. New York, NY: Penguin.
8. "Margaret M. Scariano". (2013, June 30). *Missoulian*. Retrieved February 21, 2019 from https://missoulian.com/news/local/obituaries/margaret-m-scariano/article_3b1c4d9a-e196-11e2-b139-0019bb2963f4.html.
9. Grandin, T., & Scariano, M. (2005, reissue). *Emergence: Labeled autistic*. New York and Boston: Grand Central Publishing.
10. Williams, D. (1992). *Nobody, nowhere: The extraordinary autobiography of an autistic*. New York: Times Books.
11. Sinclair, J. (2010). Being autistic together. *Disability Studies Quarterly, 30*(1). Retrieved from http://www.dsq-sds.org/article/view/1075/1248.
12. Sinclair, J. (1992). Bridging the gaps: An inside-out view of autism (or, do you know what I don't know?). In E. Schopler & G. B. Mesibov (Eds.), *High-functioning individuals with autism*. New York: Plenum Press. Retrieved from http://jisincla.mysite.syr.edu/bridgingnc.htm.
13. Sinclair, J. (1999). *Why I dislike person first language*. Retrieved from http://www.larry-arnold.net/Autonomy/index.php/autonomy/article/view/OP1/html_1.

Open Access This chapter is licensed under the terms of the Creative Commons Attribution 4.0 International License (http://creativecommons.org/licenses/by/4.0/), which permits use, sharing, adaptation, distribution and reproduction in any medium or format, as long as you give appropriate credit to the original author(s) and the source, provide a link to the Creative Commons license and indicate if changes were made.

The images or other third party material in this chapter are included in the chapter's Creative Commons license, unless indicated otherwise in a credit line to the material. If material is not included in the chapter's Creative Commons license and your intended use is not permitted by statutory regulation or exceeds the permitted use, you will need to obtain permission directly from the copyright holder.

3

From Exclusion to Acceptance: Independent Living on the Autistic Spectrum

Martijn Dekker

In 1985, as an oblivious and undiagnosed autistic 11-year-old with no idea who I really was or what my life was for, I was introduced to the amazing and captivating world of programmable home computers, who always mean what they say and say what they mean, and expect nothing less of you. Thus I acquired my most central and enduring identity, that of a computer programmer. It was then that my real social life began.

Though I live in the Netherlands, I lived most of that real social life in plain text, worldwide, in English, over a metered telephone line and a modem, at a nice and safe distance from my daily worries. The Internet was still far from accessible to mere mortals, so I used dial-up hobby computer systems called BBSs, bulletin board systems. These BBSs "echoed" personal and group email to each other by exchanging it nightly over the phone during cheap hours, thus forming a slow-lane network called Fidonet.

M. Dekker (✉)
Groningen, The Netherlands
e-mail: martijn@inlv.org

© The Author(s) 2020
S. K. Kapp (ed.), *Autistic Community and the Neurodiversity Movement*,
https://doi.org/10.1007/978-981-13-8437-0_3

Eventually, I became the moderator of two worldwide Fidonet echomail conferences.

While this did gave me a sense of accomplishment, I was living what others call 'real life' without a diagnosis or a clue. As I was growing into a young adult, I deteriorated. Along with depression, I developed what I now know to be catatonia-like difficulties with self-direction, such as taking any sort of initiative [1]. The depression went away later, but that mysterious and near-total inability to "just do it!" never has.

In 1995, as my education and my offline life had fallen apart completely, the Internet became accessible to the common people. On it, my mother and I discovered Asperger's syndrome. Less than a year later, I had a diagnosis of 299.00 Autistic Disorder by DSM-IV (Diagnostic and Statistical Manual of Mental Disorders, [2]) criteria.

With nothing left to lose but my family's love and support, and on a self-discovery high, I went out into the new world of the 'Net. At the time, it was nearly as textual as BBSs, but I was enthralled to find that worldwide communication was instant instead of taking days. I found a few mailing lists (email-based discussion groups) dominated by parents of autistic children and professionals, plus one managed by autistic people: ANI-L, the list run by Autism Network International (ANI), founded in 1994 and hosted by Syracuse University since 1996 [3].

By the time I entered the scene, ANI had developed a vibrant and specific autistic subculture, with verbal stims (in plain text) and various in-jokes involving fipples, llamas, and the like. It was quite wonderful to see this working so well for those who fit in. Here were, after all, people proving for the first time that autistic community is possible! But their ways were not every autistic person's cup of tea, and their community did not seem all that welcoming to those with dissenting opinions. The need for an alternative was somewhat apparent.

But in this pre-social media era, with the World Wide Web still a toddler, starting a group of any description on the Internet was non-trivial at best. No services existed that let you do this at the click of a button. It required the use of server software that usually was very expensive and always needed to be managed by an expert. Nor did I have connections to a university or corporation that could host it for me. So, although there was just the one autistic community, the idea of starting my own didn't occur to me.

Then I was contacted, in early 1996, by the American father of an autistic son, who was advocating for another American autistic man who I shall refer to as X. With a decades-long history of trauma, X had serious behavior problems. In part, this was a conscious decision: he was fed up with being mistreated, so, in his words, "no more Mr. Nice Guy". But it was also clearly the effect of a long history of institutionalization.

So he found himself banned from the few autism communities that existed on- and offline, including the only autistic-run one. But some people went much further: they actively spread the image of him as a direct threat, seriously and physically dangerous, apparently to ensure he would never get another chance elsewhere.

To me it was obvious that traumatized autistic people frequently have problems with bad behavior. Surely, effective advocacy would require at least making an effort to include the really difficult ones, too? It physically hurt to see some of my fellow autistic people not only eject, but also go out of their way to further damage someone who was already broken. My reaction, evoked in part by my own traumatic memories of schoolyard violence, peer rejection, and paternal authoritarianism, was visceral. I felt and understood this man's anger and desperation. Empathy! Much as I disapproved of X's methods, this reaction to them seemed so much more harmful that, despite my usual inertia, I found myself spurred into action.

Starting an Internet group seemed impossible without connections to a large organization, so I suggested that X try starting a BBS of his own, of the old-fashioned dial-up variety. He had some relevant experience and I figured that it might help him to bear the responsibility himself, to achieve something that was his own. Aiming it at people who had been "shunned, hated or misunderstood" due to being autistic or otherwise different, he set up forums on various relevant topics, with a definite activist bend. It was not networked, so I dialed in to the USA to participate at considerable expense. The BBS never had more than five users, and after a month or so, even they stopped coming. Whatever else may have contributed to the lack of interest, it was clear that with the Internet growing explosively, BBSs were on their way out.

Then, serendipity intervened. A new kind of Internet provider opened in the Netherlands, a subsidiary of a UK provider oddly named Demon Internet, after the English expression "to work like a demon" with a nod to "daemons" as in server software on Unix and similar operating systems. They were special in not only technically enabling, but actively encouraging their users to run server software—to become a full-fledged host on the 'Net alongside giants like Syracuse University. Freeware mailing list server software called Macjordomo had also become available. Power to the people! The puzzle pieces fell into place.

I offered X the position of co-moderator, but at this point he had lost interest in taking an active role. So he took a back seat as I started it as my group. As was customary, the BBS had been divided into various topic-based forums; I decided to keep that aspect by creating several mailing lists dedicated to similar topics. I began to operate these mailing lists together as a set, so members could choose their topics of interest in which to participate.

Thus, the first entirely self-run and self-hosted autistic community on the Internet was born in July 1996, called Independent Living (InLv), with the Internet hostname of inlv.demon.nl (later changed to inlv.org). From then on, until cable internet became available in 2000, I had a routine of actively distributing group mail over the Internet through my dial-up line in batches, a few times a day. The communication was slower than ANI-L, but still much faster than Fidonet.

The group grew quickly. While some members were non-autistic friends and sympathizers, often with other neurological conditions, most of us were autistic—some recently diagnosed as adults, others seeking and receiving diagnoses as a result of their membership, still others content with having self-identified. In finding each other, we found ourselves. The collective process of self-rediscovery as autistic people that we went through as members was so intense, I stopped engaging in computer programming altogether for a number of years. I made friends, and more friends through friends, in various countries. Thus, after that of a computer programmer, I acquired my second-most central and enduring identity: that of an autistic person. It was then that I felt truly accepted in a community for the first time.

The text-only email nature of our community, far from being limiting or disabling, was found to be an advantage. We were able to skip all the social rituals and awkwardness and cut right to the chase, undistracted by body language, timing, sensory or eye contact issues, or any of the other autism-related difficulties with socializing. In the words of one member, the Internet was where "people can see the real me, not just how I interact superficially with other people" [4]. This helped us support each other more effectively. Email seemed like a natural communication medium for us autistics, like sign language is for deaf people [5].

The notion that we lack empathy was quickly deconstructed as it became clear that neurologically typical (NT) people had considerably less empathy with us than we had with them. A lot of lifelong pain was shared and empathized with. As part of processing those experiences, we started developing our own theories of neurotypicality—of why these strange people, who form the majority, do what they do. We had a bit of fun with it; tongue-in-cheek terms like "neurotypical syndrome" and "social dependency disorder" were thrown around. Some of us also felt inspired to explain ourselves to the neurotypical population using our newly found collective insight [6]. As we were so used to being misunderstood, patronized, and pathologized, it was a relief to have the shoe on the other foot.

But we also started finding things in common with each other, things that were not part of the diagnostic criteria for autism. For instance, many of us had trouble recognizing faces to various degrees, relying on other features such as voice, gait, and hair and clothing style to recognize a person. One InLv member, Bill Choisser (1947–2016), found an obscure medical term for this condition: prosopagnosia. It was thought to be very rare, but the group's experience clearly suggested otherwise. Bill popularized an easier to use term for it: face blindness. Based in large part on InLv discussions, he wrote the first book about the condition, which he published online [7]. The book gained traction and spread knowledge on the condition, and the term is now widely used.

Meanwhile, as people in and (mostly) out of the Netherlands began to take note of my activities, I began to be invited to autism conferences as a speaker. I would warn the InLv members of the impending silence when I was about to travel, then upon arrival I would find places to plug my laptop into a phone line and distribute the backlogged mail.

The feeling, at once grave and uplifting, of having an entire worldwide human community inside your laptop computer, depending on your own continued action to survive, is hard to describe. Wherever I went, they went with me. As I boarded airplanes, the announcement "in the event of an emergency, you must leave all hand luggage behind" acquired an existential level of fearsomeness.

In July 1997, when InLv was a year old, I met X in person in Minneapolis, Minnesota, USA as the two of us were invited to speak on the new phenomenon of online autistic community and activism at a conference organized by the Society for Disability Studies. In person, he seemed much more timid than intimidating. The conference speech was a success for both of us.

But on InLv, though X continued to prefer his back seat, he was growing increasingly frustrated with the members' self-discovery and mutual support, which he had begun to see as spineless psychologizing. He wanted action. The rest of us were not ready for action. He also became increasingly jealous as I gravitated toward some level of prominence in certain autistic circles. It's understandable: where he had failed, I was succeeding.

Nevertheless, his behavior deteriorated to the point where it became both detrimental to the group and personal to me. Mere weeks after our conference speech in Minneapolis, I had to remove him. He and I both know the exact reasons, and that is enough. I have avoided contact since, but I wish him well. The group had succeeded in including him for a full year. In the process I had learned some hard lessons about both the possibilities and the limits of inclusion, which proved invaluable in later years.

InLv continued without him just as the discussions had begun to gravitate from the purely personal to the more political. A new idea came up in the group, based on the evidence and lived experience that autistic brains are wired differently from the mainstream on a fundamental level. Biological diversity of all kinds is essential to the survival of an ecosystem—so why should neurological diversity, which is one aspect of biological diversity, be any different? The objective fact that neurological diversity exists emerged as a strong argument for the acceptance of autistics and other neurological minorities as distinct classes of people among many, who have something valuable of their own to contribute, and who are as inherently worthy of equal rights as anyone.

In 1998, Judy Singer from Australia, who identified as having "AS [Asperger's Syndrome] traits", turned these InLv discussions into an influential sociological thesis [8] and book chapter [9], citing plenty of group members with their permission, and adding the requisite academic language to lend it legitimacy. Thus, she is correctly credited with coining the term 'neurodiversity' [10]. However, it may be argued that the American journalist Harvey Blume, who was also an InLv member and whom Singer cites as a frequent discussion partner, first popularized the term [11]. What is certain to me is that InLv, due to the ethos of acceptance, inclusivity, and rejection of social and political conformism that I imparted on it, was able to provide the environment in which the idea could emerge.

It is important to note that InLv's notion of neurodiversity was different from the "neurodiversity paradigm" that many contemporary activists subscribe to. These days it is often held that there is no such thing as a brain that is "less" or "broken" because "all neurologies are valid" [12]. By contrast, neurodiversity as an aspect of biodiversity includes and accepts people with suboptimal neurological configurations. While autistic people who would have preferred to be "cured" if possible were a minority in the InLv community, we never excluded or denounced them.

Meanwhile, the InLv community was joined by the #asperger IRC (Internet Relay Chat) channel for which I took over management in 1998. It had been started in 1997 by a German man nicknamed Nox, who had a diagnosis of schizoid personality disorder. He created the channel to be only for people on the autistic spectrum and related conditions. I disagreed with the exclusion of neurotypical guests and still do, but I did not feel like I could change this after the channel had become established as what it was. In any case it provided a way for autistic people, including many InLv members, to have text-based conversations that are much more direct than email.

Soon, the combination of these two communities started carrying over into the physical world. Many "real-world" relationships resulted, and I would estimate that at least a dozen children were born because of them, including my own three. Amazingly, the #asperger channel survives to this day, though people who join need to be patient as activity is intermittent.

Around the turn of the century, my own catatonia-like inertia problems started affecting my ability to manage the group effectively. As my initial burst of initiative petered out, it became harder and harder for me to manage email requests in a timely manner, and new members had to remind me multiple times and wait months before being added. Some never received a response, and thinking of that fills me with guilt to this day. Thankfully, as easy-to-use mailing list and forum hosting services became available on the web, other autistic-run communities started popping up. In spite of all this, InLv continued until early 2013. Sixteen years is a good run for any online community.

Probably the most significant real-life outgrowth of InLv and related communities is the yearly Autscape residential conference (see Buckle, Chapter 8), founded after a 2004 InLv discussion on the idea of creating a European equivalent of Autreat. As one of the Autscape organization's directors, it makes me happy to see InLv's spirit of inclusion and acceptance continue there.

References

1. Wing, L., & Shah, A. (2000). Catatonia in autistic spectrum disorders. *The British Journal of Psychiatry, 176*(4), 357–362.
2. American Psychiatric Association. (1994). *Diagnostic and statistical manual of mental disorders* (4th ed.). Washington, DC: American Psychiatric Association.
3. Sinclair, J. (2005, January). *Autism network international: The development of a community and its culture*. Retrieved from http://www.autreat.com/History_of_ANI.html.
4. Blume, H. (1997, June 30). Autistics, freed from face-to-face encounters, are communicating in cyberspace. *The New York Times*. Retrieved from https://www.nytimes.com.
5. Dekker, M. (1999). *On our own terms: Emerging autistic culture*. Autism99 online conference. Retrieved from http://www.autscape.org/2015/programme/handouts/Autistic-Culture-07-Oct-1999.pdf.
6. Meyerding, J., KB, Clark, P., & Comm, M. (1998). *Why are we so unfriendly? Or: Hello friend, now please go away* (Web log post). Retrieved from http://www.inlv.org/subm-social.html#unfriendly.
7. Choisser, B. (1997). *Face blind!* Retrieved from http://www.choisser.com/faceblind.
8. Singer, J. (1998). *Odd people in: The birth of community amongst people on the autistic spectrum: a personal exploration of a new social movement based*

on neurological diversity (Honours dissertation). University of Technology, Sydney. Republished in Singer (2017).
9. Singer, J. (1999). 'Why can't you be normal for once in your life?' From a 'problem with no name' to the emergence of a new category of difference. In M. Corker & S. French (Eds.), *Disability discourse* (pp. 59–67). Milton Keynes, UK: Open University Press.
10. Singer, J. (2017). *NeuroDiversity: The birth of an idea*. n.p.: Judy Singer.
11. Blume, H. (1998, September). Neurodiversity: On the neurological underpinnings of geekdom. *The Atlantic*. Retrieved from https://www.theatlantic.com.
12. Grace, A. (2015, February 24). *Ten things you reject by embracing neurodiversity* (Web log post). Retrieved from http://respectfullyconnected.com/2015/02/ten-things-you-reject-by-embracing.

Open Access This chapter is licensed under the terms of the Creative Commons Attribution 4.0 International License (http://creativecommons.org/licenses/by/4.0/), which permits use, sharing, adaptation, distribution and reproduction in any medium or format, as long as you give appropriate credit to the original author(s) and the source, provide a link to the Creative Commons license and indicate if changes were made.

The images or other third party material in this chapter are included in the chapter's Creative Commons license, unless indicated otherwise in a credit line to the material. If material is not included in the chapter's Creative Commons license and your intended use is not permitted by statutory regulation or exceeds the permitted use, you will need to obtain permission directly from the copyright holder.

4

Autistic People Against Neuroleptic Abuse

Dinah Murray

Origins

In the mid-90s I had a job as a support worker for people with severe and multiple learning (intellectual) disabilities including autism, who had been discharged from National Health Service (NHS)-run long-stay hospitals into "the community." Our training for this job was thorough and humane in many ways, but little was said about the stacks of ring-bound blister packs of pills kept under lock and key. A community pharmacist explained to us how to administer the pills in the right order at the right time with proper regard to hygiene and record keeping. Nothing was more important, it seemed.

At first I took it for granted there were good reasons for all the pills: after all, these people had come out of hospitals. However, after a while I noticed that thioridazine was one of the "medications" and it rang a bell as one of the "old, dirty" antipsychotics aka "major tranquilizers"—well-known to cause severe movement disorders *inter alia*. I started looking up

D. Murray (✉)
London, UK

the other drugs, then someone left this photocopied article in the office about the use of antipsychotic drugs with adults with learning disabilities and challenging behavior. These disabled people were being deliberately sedated with extremely harmful drugs.

One night the young manager, in tears, told me she'd just discovered that one of our most severely disabled people, who had gone into the old institution when he was 4 or 5 years old, had *ridden his tricycle into the long stay hospital where he had then lived—and been drugged with major tranquilizers—for 40 years*. We knew him as someone who needed help even to turn over. The period when our people had been hospitalized was a period of excited experimentation with new drugs and the doctors really had no idea what they were doing when they prescribed doses of chlorpromazine (an antipsychotic) to children which would later be regarded as around twelve times the recommended maximum. I was in tears too.

I became obsessed with the medication. It was increasingly clear that we were looking at routine, unquestioned, administration of substances that everybody in the know knew to be dangerous, to people that everybody in the know knew to be powerless: the rage drove work lasting into the twenty-first century.

As I got deeper into the general research I was doing, I became more and more horror-struck by the great range of problematic features of the whole psychotropic prescribing business, which were stripped bare in the power structures in which people with no information and no voice were ultimately vulnerable. I discovered that the drugs had a negative impact on hormones, insulin, dopamine, teeth, all drives including sex drive; I discovered these drugs kill people as well as twisting their limbs; I discovered that there was a long-term cumulative harmful impact and that the movement disorders many develop include a terrible restlessness, emotional and physical, as well as catatonia-like loss of function and uncontrollable tongue and limb movements; that polypharmacy—multiple prescribing—was often to counter iatrogenic impacts of other drugs; that it had long been known that previously sane monkeys had gone bonkers after being dosed with a neuroleptic for several months which was suddenly withdrawn, causing not "relapses" but discontinuation syndrome, and gradual withdrawal was strongly recommended but rarely followed; that people were being medicated for distressing events and their signs of misery were seen as challenging instead of them being helped to deal with their understandable grief and upset; that urinary incontinence was a possible adverse effect (with huge social consequences) that was regarded as insignificant; that lactation was sometimes triggered, even in males; that weight gain to the point of obesity was shrugged off in the research but punished in the kitchens by stricter diets and locked cupboards. Worse still, all claims of "successful outcomes" were based on reductions in behavior of various sorts—given the usual sedating effects of the major tranquilizers, their ability to reduce behavior was unimpressive—the sedative effect wears off after 6–8 weeks, few studies at that time lasted longer than 6–8 weeks (see Fig. 4.1).

So I left my regular job and signed on as a relief worker, a casual worker status that meant I could never be asked to distribute medications. This also gave me more time for research and to develop a campaigning website. Thanks to pharmacologist Paul Shattock who told me about it—he was publicly concerned about the use of neuroleptics for behavior control long

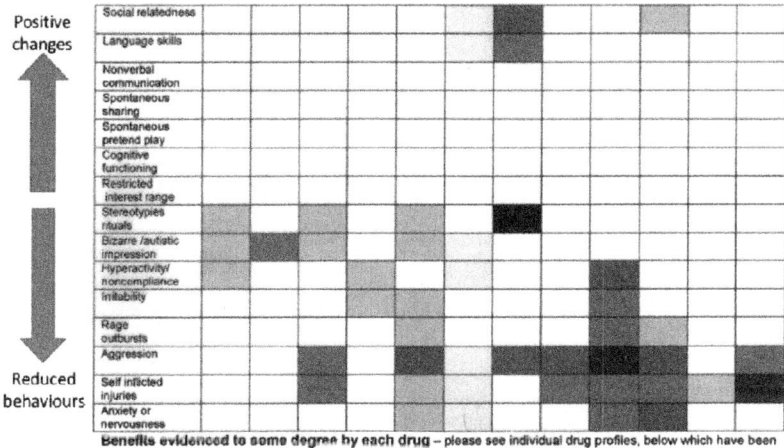

Fig. 4.1 The chart clearly showed that the vast majority of claimed "benefits" of medication in this sphere are about reducing behavior rather than enhancing personal well-being or capacity

before his public support for Andrew Wakefield's dubious research into the MMR vaccine—I went to an Autism Europe Conference in Brussels and heard about their Code of Good Practice on Prevention of Violence against Persons with Autism [1]. Paul also introduced me there to a concerned mother from the UK; meeting her at that conference was the first necessary step toward getting some activists working together in the tiny and unstructured group we called Autistic People Against Neuroleptic Abuse, or APANA for short.

The point of APANA was to be an effective vehicle to raise awareness of the harms being done to vulnerable people in the name of care, and to penetrate some entrenched positions in huge and deep-rooted power structures. For people with access (not the subjects of my case studies!), the Internet proved a rich source of information about both government and NGO thinking on these and related issues: there were consultations, and guidelines, and that was one way to get one's voice heard.

Getting to Work and Forming Alliances

APANA recruited two autistic people: as Chairperson, David Andrews, who was in the process of acquiring several psychology-related qualifications; as Patron, Wendy Lawson (now known as Wenn Lawson), who was well along a similar path. The rest of the team were all parents of adult offspring they had seen damaged by psychiatrictreatment. I was spending more and more time reading research (for example, [2–6]). I discovered that an unexpected result of the hospital closures was that prescriptions of psychotropics had gone *up* as neighbors of the new noisy people in the community complained. When I looked into the old prescription records, I found that this was true for almost everyone in the houses where I most often worked. The institutions had eventually had a policy of reducing medication.

There was one discovery in the files that tipped my concern from a commitment to an out-and-out mission. A service user I'll call Patrick, who was on the highest neuroleptic dose I found in this group, had become borderline catatonic. From the files it was clear that he had once been fairly lively. Then I found a letter, addressed to senior management from the "community" learning disability psychiatrist. She complained in her letter that support workers had questioned her judgment that Patrick's medication should be increased by 30%. Since the rules that govern care providers explicitly require them to follow all medical advice or put their business at risk, this reproach was significant and probably got some caring and conscientious workers into a bit of trouble. Much worse though were the repercussions for Patrick: the psychiatrist had increased his dose by 50% after her authority had been questioned in this outrageous way; without prompting he was no longer able even to complete the action of putting a kettle on to boil. It is ironic that "health and safety" risk aversion in social care settings led to routine, authorized, high-risk behavior by both staff and management toward the people receiving their "care."

I discovered that distinguished psychiatrist Lorna Wing was interested in autism and catatonia, and had expressed concerns that neuroleptic prescribing was sometimes implicated in its onset. So I rang Lorna Wing's Centre, then known as the Centre for Social and Communication Disorders, in the hopes of speaking to her. She was not there that day but Judith Gould came to the phone to deal with this unknown support worker's anxious query. She listened to the problem, immediately said I should ring Lorna Wing herself and gave me her home phone number. So next I picked up the phone and dialed that. The great Wing answered the phone herself and was so interested and open that I trusted her at once and said very soon after our conversation began, that all I had found among her colleagues was "Arrogance, ignorance, and hypocrisy"—to which, taking my breath away, she replied with vigor, "I couldn't agree more!" One couldn't hope for a better ally in this particular battle [7].

Communicating Our Message Across Many Platforms

I decided to write up case studies of what I'd found in the files in a way that would make the research as effective as possible. I looked at quality of life issues and the impact on those of the ramified adverse effects of the prescribed drugs; I compiled detailed timelines for four service users, three of them autistic, and their life events [8].

Unfortunately, how to carry out quality of life assessments objectively remains a vexed question to this day [9], so instead I took and followed advice on assessing relevant costs, such as travel and staff time, in British pounds and went back through the records adding actual costs or rule-governed estimates of costs to the timelines.

Our website, run by my friend Sue Craig, had much factual information, including all the illustrations to this chapter, and useful links and ancient advertisements for old drugs and new. Canadian artist Ralph Smith designed an elegant logo for us (Fig. 4.2). This served us well and got the message out that Autistic People Against Neuroleptic Abuse was an active force. Autistic activists in the USA such as A. M. Baggs (now

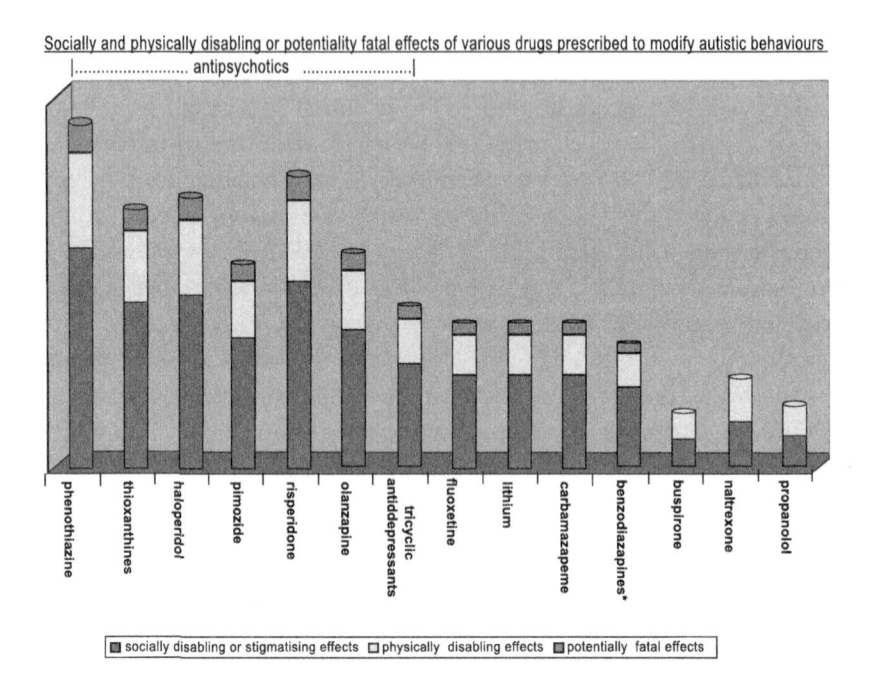

Fig. 4.2 Summarized comparisons of restrictive or potentially fatal effects of psychotropic drugs prescribed for autistic adults with learning disabilities and challenging behavior (ca 2001). From left to right, starting with the antipsychotics: phenothiazines (e.g. Chlorpromazine), thioxanthines (e.g. Fluenthixol), Haloperidol, pimozide, risperidone, olanzapine; and then various non-neuroleptic experimental drugs for autism-related perceived problems, viz paroxetine, lithium, carbamazepine, buspirone, naltrexone, and the beta-blocker propranolol

known as Mel Baggs) and Kassiane Sibley (now Kassiane Asasumasu) were supportive and deepened my understanding of what it is like to be on the receiving end of interventions designed solely for the purpose of suppressing behavior which other people condemn.

Being able to brand the work as from "APANA" was I think particularly helpful in being taken seriously rather than assimilated into the *vox populi*. Mencap, the main British learning disabilities charity, agreed to circulate the research to their consortium of service providers in this field.

It also helped that I had some strategically placed friends and allies. David Branford, a senior learning disabilities pharmacist, was sympathetic, and encouraged me to attend a conference he was organizing in Leicester late last century. It was there I first encountered the "psychiatry is to real medicine as astrology is to astronomy" meme, inadvertently shared with me by a psychiatrist who misjudged my status until rather late in our conversation, assuming I was a fellow clinician.

Rita Jordan at Birmingham University gave me a platform for the medication issues on one of the Autism Distance Education Course weekends, thus reaching everyone doing the course at that time. Almost all of those were professionals in the field, some in senior positions, including people who worked with adults. Glenys Jones was in the process of setting up a new, practical, autism-relevant journal, *Good Autism Practice*, and she invited me to submit my research for the prototype issue, published in 1999. I also got a poster presentation at the Autism Europe Conference in Glasgow that year and Wen Lawson and/or I were there in person throughout ready to discuss it.

Pressing Parliament and Leaning on the Law

I undertook a careful analysis of a range of medications proposed at the time for "ameliorating autistic behavior," and I scored adverse effects according to their recognized frequency (using the free Medline database). Risperidone, newly popular with prescribers, was only slightly less harmful than chlorpromazine at the recommended (like-for-like) dose ranges for psychosis (see Fig. 4.3). Wenn Lawson and myself addressed a sub-committee of the All-Party Parliamentary Group

Fig. 4.3 The APANA logo, designed by Ralph Smith

on Autism about prescribing, illustrated with Fig. 4.1. (One of the people who heard us was Virgina Bovell, a mother of an autistic person with learning disabilities, in whom we found a new ally despite her involvement with behaviorism—I later came to understand that parents may be presented with behavioral approaches as the only alternative to a medicalized attitude and recourse to drugs.)

I got views from as many as possible autistic people with relevant experience and discovered that some, but not all, were saying that at a very low dose, they found risperidone positively helpful; that it improved their mood and could make social encounters less stressful. If someone tells me that they find a drug helpful and their consumption is moderate, I'm going to see that as normal human practice. In the case of risperidone or other antipsychotics, I liken this to accepting that a glass of wine every day may do a lot of good, while a bottle will not. To me it is the absence of choice, the absence of relevant information, and the inability to refuse that need fixing. In the long run, it is worth noting that most of the people I knew who liked risperidone at a low dose eventually developed adverse effects that put them off—they were fortunate to have the capacity to articulate their problems and the autonomy to make this decision.

(Having now, two decades later, read through extensive messages posted in the last few years on mental health forums, it is clear that low dosing with risperidone has become commonplace, especially for anxiety. It is also clear that there is a major division in people's experiences of this drug, with some people greeting its effects with joy and others with real horror. To give something as powerful and unpredictable as this to people who are unable to tell you how they feel, still seems to me the height of irresponsibility. People need to understand the potential risks and freely choose to take them. Clearly that did not apply to the people in my case studies!)

Though our message was being heard in some quarters, the extreme power imbalance sometimes seemed overwhelming. So we decided to keep pushing positive proposals tied to precisely referenced and cited government papers, by addressing civil servants, administrators, and Members of Parliament about the law with some specific suggestions that I set out at length. These are to be found following the main text of my "Potions and Pills" piece [8] in the longer version online.

A related line I decided to follow was to get a proper legal view *pro bono* (for no fee because for the public good) if possible. Somehow I found a deeply committed solicitor, Karen Ashton, and barrister Paul Bowen of the Doughty Street Chambers in London, who were willing to look to assess the legal situation vis-à-vis medical treatment that flouts the Hippocratic Oath, Do No Harm—especially to people deemed to be "mentally incapacitated." They needed someone to gather the evidence together and I gave up my job entirely for three months, during which I created a compendium of abstracts substantiating a great variety of adverse effects and at the same time demonstrating that reduced behavior was the key index of "efficacy" in their use (see Fig. 4.1). This resulted in lawyers Bowen and Ashton producing a discussion paper (available in the online edition of this chapter), which suggested that medical treatment can amount to an assault unless great care is taken regarding consent or "best interests." They proposed that the Human Rights Court "may well be willing to exercise its power in relation to the prescribing of psychotropic drugs, particularly where serious side-effects are well-established." This was circulated widely.

In 2001, Ashcroft and colleagues [10] called for better research into antipsychotic prescribing for "challenging behaviours." They cited Brylewski and Duggan's [11] *Cochrane Review*, as showing "over 500 citations assessing the impact of antipsychotic drugs on challenging behaviour. Of these only three were methodologically sound randomised controlled trials, but even these were unable to show whether antipsychotic drugs were beneficial or not in controlling challenging behaviour" [10].

Impact

Ashcroft and colleagues frame the issues thus: "People with learning disability sometimes display challenging behaviour. This can be managed by use of antipsychotic medication or behavioural therapy or both. There is no solid evidence, however, that these therapies are safe and effective." Unfortunately the possibility that behavioral therapy may not be safe was not pursued, nor was the possibility that a focus on behavior control cannot preserve mutuality, create trust, or be authentically "person-centered." This fixation with behavior, along with skillful marketing of "Positive Behaviour Support," has underpinned and undermined a medical campaign against the drugging launched in 2018 (see below).

Consent issues and the best interest concept were soon to be leading themes in the Mental Capacity Act (2007), (a development to which Paul Bowen contributed). That is a very strong piece of rights legislation in principle, though it has thrown up some paradoxes in practice (see, e.g., DoLS discussion at House of Lords 2015, or this Parliamentary video from 2018, https://www.parliamentlive.tv/Event/Index/d47bf41e-72b1-48d8-afc5-b5727a40f05b). The MCA guidelines draw attention to the possibility that the psychotropic effects of some medications may hinder judgment, and there are widespread guidelines on administering medication to people whose best interests must in law be factored in. In 2006 the University of Birmingham published an attempt to address the consent issues by creating a simplified symbol-based system for describing the medications and their effects [12]. Perhaps our activities contributed to the wider recognition of such needs, but how much long-term impact did we have?

Maybe a bit for a while—however, see this from the Foreword of the Faculty of Learning Disabilities of the Royal College of Psychiatrists [13]:

> There is compelling evidence that a significant number of people with intellectual disabilities are prescribed psychotropic medication that, at best, is not helping them. In particular, there is a risk that doctors are prescribing medication to treat behaviour that is an expression of distress or a mode of

communication rather than a mental disorder. Some people with intellectual disabilities have difficulty communicating their emotional needs and preferences. Therefore, doctors have a particular responsibility to ensure that they have fully assessed a person's potential to benefit from medication before they prescribe. They must also check that the anticipated benefits have occurred after they have prescribed.

David Branford, whose earlier work [5] influenced mine, co-edited this careful and strongly worded document produced by a team dominated by learning disability specialists.

With two provisos, this specialist report's advice is generally clear and strongly argues its case for greater prescribing caution. One reservation is that the advice lumps in autistic people with all other learning disabilities and generalizes that usual dosing practices will be fine for addressing mental illness when it occurs. Much anecdotal evidence says that autistic people often have atypical reactions including super sensitivity to drugs. As Defilippis and Wagner [14] suggest, "Children and adolescents with autism spectrum disorder appear to be more susceptible to adverse effects with medications; therefore, initiation with low doses and titrating [adjusting the dosage] very slowly is recommended." Also, neither they nor the Care Quality Commission (a public regulator of health and social services in England) note the need for staff medication training. The latter says (2017) "We will not consider it to be unsafe if providers can demonstrate that they have taken all reasonable steps to ensure the health and safety of people using their services and to manage risks that may arise during care and treatment." Since "reasonable steps" will of course include following instructions from doctors, this regulation can only have a protective impact if the doctors are also changing their practices: perhaps the current alliance between the NHS and the behaviourists (the autism and learning disability campaign Stopping Over Medication of People, or STOMP) may have the power to change those.

Sadly, it seems there has been little or no real progress this century—yet the very existence of these reports shows that the zeitgeist may finally have penetrated Bedlam. We may have helped let it in.

References

1. Autism Europe. (1998). *Draft code of good practice on prevention of violence against persons with autism.* Brussels: Autism-Europe.
2. Kiernan, C., Reeves, D., & Alborz, A. (1995). The use of anti-psychotic drugs with adults with learning disabilities and challenging behaviour. *Journal of Intellectual Disability Research, 39*(4), 263–274.
3. Shiwach, R. S., & Carmody, T. J. (1998). Prolactogenic effects of risperidone in male patients—A preliminary study. *Acta Psychiatrica Scandinavica, 98*(1), 81–83.
4. Manchester, D. (1993). Neuroleptics, learning disability, and the community: Some history and mystery. *BMJ, 307*(6897), 184–187.
5. Branford, D. (1996). Factors associated with the successful or unsuccessful withdrawal of antipsychotic drug therapy prescribed for people with learning disabilities. *Journal of Intellectual Disability Research, 40*(4), 322–329.
6. Tranter, R., & Healy, D. (1998). Neuroleptic discontinuation syndromes. *Journal of Psychopharmacology, 12*(4), 401–406.
7. Wing, L., & Shah, A. (2000). Catatonia in autism spectrum disorders. *British Journal of Psychiatry, 176,* 357–362.
8. Murray, D. (1999). 'Potions, pills and human rights' in opening volume of *Good Autism Practice* (long version). Retrieved from http://www.autismusundcomputer.de/potions.en.html.
9. McConachie, H., Mason, D., Parr, J. R., Garland, D., Wilson, C., & Rodgers, J. (2018). Enhancing the validity of a quality of life measure for autistic people. *Journal of Autism and Developmental Disorders, 48*(5), 1596–1611.
10. Ashcroft, R., Fraser, B., Kerr, M., & Ahmed, Z. (2001). Are antipsychotic drugs the right treatment for challenging behaviour in learning disability?: The place of a randomised trial. *Journal of Medical Ethics, 27*(5), 338–343.
11. Brylewski, J., & Duggan, L. (1999). Antipsychotic medication for challenging behaviour in people with intellectual disability: A systematic review of randomized controlled trials. *Journal of Intellectual Disability Research, 43*(5), 360–371.
12. Unwin, G., & Deb, S. (2006). *Your guide to taking medicine for behaviour problems: Easy read.* University of Birmingham. Retrieved September from https://www.birmingham.ac.uk/Documents/college-les/psych/ld/LDEasyReadGuide.pdf.

13. Alexander, R., Branford, D., & Devapriam, J. (2016). *Psychotropic drug prescribing for people with intellectual disability, mental health problems and/or behaviours that challenge: Practice guidelines.* (Faculty Report No. FR/ID/09). Retrieved from The Royal College of Psychiatry website: https://www.rcpsych.ac.uk.
14. DeFilippis, M., & Wagner, K. D. (2016). Treatment of autism spectrum disorder in children and adolescents. *Psychopharmacology Bulletin, 46*(2), 18–41.

Open Access This chapter is licensed under the terms of the Creative Commons Attribution 4.0 International License (http://creativecommons.org/licenses/by/4.0/), which permits use, sharing, adaptation, distribution and reproduction in any medium or format, as long as you give appropriate credit to the original author(s) and the source, provide a link to the Creative Commons license and indicate if changes were made.

The images or other third party material in this chapter are included in the chapter's Creative Commons license, unless indicated otherwise in a credit line to the material. If material is not included in the chapter's Creative Commons license and your intended use is not permitted by statutory regulation or exceeds the permitted use, you will need to obtain permission directly from the copyright holder.

5

Autistics.Org and Finding Our Voices as an Activist Movement

Laura A. Tisoncik

Deep Origins: Martin Luther King Jr. and the Fair Housing Campaign

Many have said that the first explicitly political act that emerged from the autistic community was the website autistics.org. A few people have said that I, as the founder of autistics.org, am the founding mother of autistic activism.

That's not true. While the community, at the time of the founding of autistics.org in 1998, was oriented toward support groups, political activism was in the air. Some persons active at that time, notably Cal Montgomery, had roots in the broader disability rights, psychiatric survivors, and developmental disability movements, and were posting cogent political positions on Listservs, Usenet, and other now almost forgotten corners of the internet before autistics.org was even imagined. Nothing arises from a vacuum, and neither did autistics.org.

L. A. Tisoncik (✉)
Burlington, VT, USA

But if it is necessary to give credit to just one individual for the founding of autistics.org, I know who that person is. The true founder of autistics.org is Dr. Martin Luther King Jr. and he founded it in Chicago in 1966.

In 1966, Martin Luther King moved to Chicago, to lead a housing desegregation campaign. Dr. King was not then seen as the harmless and mostly fictional figure we venerate today. He was a radical agitator, hated and feared by most white people, and considered a criminal and likely communist by the government. The difference between MLK Jr. and more "radical" figures like Malcolm X and the Black Panthers was one of tactics, never of content, and no one during his lifespan doubted or watered down his militancy.

I was nine years old. The Chicago neighborhood I came from, Marquette Park, and the suburb we'd recently moved to, Evergreen Park, were targeted for desegregation marches that summer. People in the neighborhoods were terrified, and many of them prepared to "defend" the neighborhoods, with rocks, bricks, and baseball bats. When the Archdiocese of Chicago announced (in response to the campaign) that it would begin busing African American students to desegregate my school in the fall, violence and threats of violence against African Americans reached near wartime intensity. A house, a few blocks away, was burned to the ground amid rumors (probably false) that it had been sold to an African American family. And gangs of teens had begun patrolling the local mall with baseball bats, seeking out African Americans who dared to go shopping in a white neighborhood.

It was a time when virtually every family watched the 6 p.m. news together, and so I knew about the civil rights movement and the growing anti-war movement. I had not yet formed an opinion about current events, or even an opinion that I should have an opinion. All of it had seemed like stories of faraway events unrelated to my life, told on a 12″ black-and-white cathode ray tube in the kitchen.

But during the tumult of that summer, a pattern was emerging. The same persons who were carrying out acts of violence against African Americans were also the worst of my bullies. The few brave families who dared to volunteer as host families for African American kids about to be bused to my school were also among the few who were kind to me, the weird crazy kid in the neighborhood. I did not yet understand much about these issues, but I knew which side my enemies were choosing. I knew, therefore, which side had to be my side.

I understand racism is not the same as bullying, any more than ableism is bullying. Most of the methods any -ism uses to maintain a power differential between the privileged and the oppressed are subtle, hard to name, and even harder to prove. Many of the methods are baked into the way things are done, so that they can't be uprooted without questioning fundamental assumptions about how the world should work. But in the end every -ism will resort to overt violence, if it must, to maintain itself. The pattern I saw as a kid came about largely because the worst people are inclined to poly-bigotry: they hate everyone not evidently of their own kind.

My Activist Past

Over the course of the next few years, activism became my perseveration and my social crutch. I was awkward and fearful to the point of panic in most social settings, but as an activist I could talk to anyone, speak to any crowd, and act fearlessly. I organized my high school's underground newspaper. I became involved with the Chicago area working-class youth movement *Rising Up Angry*. I read book after book of left-wing theory, identified as an anarcho-syndicalist,[1] and joined the radical syndicalist union, the Industrial Workers of the World. I protested against military recruiters at my high school, I distributed leaflets for the defense committee for the imprisoned African-American anarchist activist, Martin Sostre. I picketed for the United Farm workers union, and I sat on the train tracks in Colorado to shut down the Rocky Flats nuclear weapons facility. I came out as a lesbian in 1974 in my first year of college, and I plunged myself into lesbian and gay politics because, even if I hadn't a clue how to meet women, I did know how to promote political causes. When I was engaged in activism, I was almost normal, or at least I was useful enough for the cause that my shortcomings could be overlooked.

---

[1] A left-wing anarchist who sees revolutionary industrial unionism as the means workers can use to overthrow capitalism.

I was less functional than frantic, less social than busy, and everything I did was as much crushing as it was a social crutch. Sometimes the stress won. Sometimes the weird screwed up everything.

I had a secret. It wasn't much of a secret, but I did my best to hide it. I was a childhood schizophrenic—a crazy person. One of Bruno Bettelheim's[2] former students had said so. It was a time I wished I could forget but never could. A very bad time, imprisoned in a very cruel place. I knew that crazy people, too, were organizing, and I supported what they were doing, but I did not want that label around my neck because I wasn't crazy, not the way they said I was. I didn't see anything or hear anything and I knew perfectly well what reality looked like. But I wasn't normal. I couldn't hold a proper conversation, or have a girlfriend, and getting a telephone call—not the content of the call, the mere fact of its ringing—could turn me into a blithering panicked mess. A lot of things could do that. I was always on the edge of panic. It was during one of my falling apart times in college, when I could not avoid the open secret that I was dysfunctional, that I came into contact with the office of students with disabilities and, from there, the disability rights movement.

Finding Out About Autism

Fast forward decades later: Though the activism of the 1960s and 1970s had faded, I was still a committed activist. Activism was no longer a crutch, but an ethical commitment, to right what was wrong and to side with the weak against the predations of the strong. I was much more content as a human being, much less stressed, though the weird was never far away. I could answer telephones though I hated them and occasionally destroyed them in frustration. I even figured out how to get into relationships, more or less, though I had no idea how to sustain one.

In the mid-late 1990s, on the early World Wide Web, search engines were absent or limited in power. It was customary to perform a search

---

[2] Bruno Bettelheim used fraudulent credentials to misrepresent himself as a psychologist and to obtain an academic position at the University of Chicago. He was exposed for plagiarism, false credentials, and abusive treatment of students after his death in 1990.

5 Autistics.Org and Finding Our Voices as an Activist Movement 69

by finding a website about your subject matter, then to follow links from website to website until you found what you were interested in.[3] What I was looking for that day was electronics parts. Several jumps into the search, I found a home page with a number of useful links, plus links to something even more desirable than electronics: data I did not know.

I love to collect information. If I let myself, I'd spend the rest of my life collecting new facts. So after I explored the links to electronics parts companies, I clicked back to the website where the author had said he had a syndrome I'd never heard of called Asperger's Syndrome. The link led to a website created by a parent of a child on the autism spectrum. I read on…

So that was what I was[4]

I plunged myself into the early autistic community, which was spread out over two mailing lists, a few Internet Relay Chat (IRC) channels, Usenet, and a scattering of websites. The focus of the early community was largely self-help. But I, like Cal before me, could not help but apply political analysis to our circumstances.

Political analysis is about power: understanding who has it, who doesn't, how the powerful have taken power over others, and how those rendered powerless can reclaim it. Once you learn how to understand injustice in political terms, you cannot help but apply political analysis everywhere you see human suffering. The skill set works nearly everywhere, because nearly everywhere, when you find a group struggling against disadvantages, you find the same dynamics.

Autistic persons are disadvantaged almost from the moment of birth. Our power to determine the direction of our lives is taken by presumptions about cognition and perception that simultaneously ignore our abilities and make unreasonable demands upon our disabilities. We are rejected by our peers, whose bullying is not merely tolerated, but encouraged, by adults, who themselves may join in the bullying. We are often rejected by our families, and many of us are murdered by them. We are placed into schools and institutions whose very purpose is to wipe us of our identity,

---

[3] The Web was a much smaller place then, so this was not as impractical as it would be today.
[4] Later I sought a professional opinion, and the professional thought my childhood development was a closer fit to PDD-NOS. According to the diagnoses of the day. Whatever. We are all Autism Spectrum now.

and whose every "treatment" and "care" is an act of violence against who we are. If we do find work, we are target number one for workplace bullying, and for being fired for autistic traits, regardless of our performance. We are first to be targeted by criminals and among the first to be targeted by police, at least in the US, where every year unarmed autistic people are among those shot by police, and where autistic people are so often stopped while going about our business we have taken to calling this the crime of "walking while autistic." We are more likely to be the victims of violence, yet we are portrayed in the media and by charities "raising awareness" as dangerous perpetrators of violence. Above all we are isolated from society at every stage as the odd, the weird, the other.

But no matter the exact life path we find ourselves on, oppression comes down to others holding or aspiring to hold undue power over us. Oppression is always the same story, and the same struggle, of the powerless against the powerful. "Injustice anywhere is a threat to justice everywhere" wrote the Great Agitator, King [1], "We are caught in an inescapable network of mutuality, tied in a single garment of destiny."

What Led up to the Creation of Autistics.Org

I joined the mailing lists and IRC channels where the early online autistic community congregated, including the main mixed parent and autistic IRC channel, #autism on Starlink IRC. There I came into conflict with a few parents over the patronizing and often demeaning way autistic adults were treated in the group (I was, after all, in my 40s, older than many of the parents who patronized me). When I circulated a petition requesting that one parent be removed as an operator for his open hostility toward autistic people on #autism, I was banned from the channel. So I started my own channel for autistics and parents, #autfriends.

#Autfriends quickly becomes a popular channel. In response the owner of #autism approached the owners of a channel for autistic persons and persuaded him to order me to close down #autfriends and tell everyone to rejoin #autism. When I refused, pointing out that he had no authority to give me or anyone else any orders outside of his own channel, and stating again that management at #autism tolerated open abuse of autistic people, the autistic channel owner banned me from his channel and mailing list, and sent the log of that private conversation to the IRC operators at Starlink IRC, possibly through the owner of #autism.

Never fear an opponent who goes too far. It is at times like this, when one's opponents overreach, that they grant you power, much as when an attacker lunges at a judoka, the force of that attack becomes the power the judo player can use to defeat the attacker. Almost every time a political action I was involved in had succeeded, it had been in part because of gross overreach by the other side.

This attack on #autfriends gave us the moral power and the outrage to create autistics.org. And of course, #autfriends reopened within minutes of the ban on Dalnet, because trying to ban an idea from the Internet is one of the more futile acts imaginable.

Sometimes It Is Allies Who Work Against You

I do not want to go into too many details about the autistic persons who worked with #autism to try to destroy us. I have no reason to stir up ancient conflicts with people who are still around and active in their own ways. No matter what the disputes are between ourselves, it's important to keep one's focus: we are not our own enemies, to be fought to the death, and no one is free from mistakes. What remains important about that incident is the truth that when one takes a stand for what is right, it is often against the objections, not just of your enemies, but of your allies.

Anyone subject to injustice is rightly fearful. It seemed to be a miracle enough that autistic adults with the means to access the net were breaking their isolation and talking to each other. The autistic community was young, small, and fragile. It is understandable why anyone who had been through what we had endured might not have wanted to take any chances, however remote, that this thin connection to humanity might be broken.

But you can't preserve a community by tolerating hostility toward it, and you can't fight injustice by succumbing to your fears. You have to learn to take a deep breath and will yourself past a pounding heart and sweaty palms to take your stand, to disrupt things as they are, and you have to have confidence in the power of truth. I had learned, over decades as an activist, to overcome fear. My objectors had little experience as activists aside from, to their credit, playing an important role in the early autistic community.

I'd been attacked more than once for my political activism. I was one of a scant dozen LGBTQ activists picketing outside a nightclub in 1976 when a much larger group of hostile counter-protesters rushed the thin police line separating us. I've had local police look for any way to arrest me on anything, real or not (I survived, but one of my fellow activists at that time was framed with heroin possession). I've been knocked down and dragged out of a public meeting by an official who did not like my political cartoons, with the culprit slamming the door shut on my arm, hard, after he was done (and then was charged afterwards, though charges were dropped when the culprit bragged about what he did to the press). I've lost a job. I can't even begin to count the number of death threats I've gotten over the years. I'd even survived an attempt to set fire to my college dorm room.

In comparison, the attacks against #autfriends were silly, and there was never a question in my mind to persist. The only strong impression they made upon me was how, if I had been seen as a middle-aged adult, gray hair and all—in other words, the way I was seen in the real world—none of it would have happened. It would be seen as an outrageous violation of boundaries to order an independent adult not to express an opinion, even in private, on the Internet, as my opponents tried to do. But I was being treated as the defective child that so many of the regulars in #autism imagined autistic adults to be, and so in their minds there were no boundaries.

Letter from a Birmingham Jail was an open letter, not to enemies of the civil rights movement, but to supporters who objected to Rev. King's tactics. "History is the long and tragic story of the fact that privileged groups seldom give up their privileges voluntarily," King wrote to his critics who had argued for negotiations instead of protests. "Individuals may see the moral light and voluntarily give up their unjust posture; but as Reinhold Niebuhr has reminded us, groups are more immoral than individuals." Politely asking the operators of the Judge Rotenberg Center to stop torturing inmates will make little headway, as all of their interests, and that of the many other centers of power that oppress autistic people, lie in the direction of continuing to do so. But picketing the home of the head of the US Food and Drug Administration to demand that shock devices be outlawed (as the disability rights organization ADAPT, in an action organized by Cal Montgomery, did recently) can begin to dislodge that evil. And before you can get to political action, you need to give voice to truths not spoken.

Autistics.Org

Autistics.org came directly out of this tumult around #Autfriends. It was envisioned as a compilation of support and resources for autistic adults, in contrast to providing resources only for parents of autistic children as #autism did. But, even as such a compilation proved too difficult for us at that time, inspired by the outrageousness of the attacks against us, speaking truths not spoken became our mission.

One of our more popular truth-telling sections was ostensibly a medical institute operated by autistic researchers, the Institute for the Study of the Neurologically Typical (ISNT). ISNT turned the tables on the dehumanization done to autistic people by autism researchers. Several contributors to that site picked apart characteristics of neurotypical individuals in the same patronizing, pathologizing, voice in which traits commonly held by autistic people are described, with feigned obliviousness to how such traits might also be useful, and no concession that neurotypicals might not be carbon-copy identical. The point of ISNT (if I can presume to speak for multiple authors here, though I do think this was a universal goal among us at that time) was to shine a light on how we are treated. Some people to this day take it literally as an assertion of autistic superiority, which leaves me wondering how deeply ingrained are their assumptions about our supposed inferiority, that they cannot recognize satire.

Understand, though, that I am not upset about claims of autistic superiority, even though I don't believe in anyone's intrinsic superiority. Every oppressed group struggles to reclaim a sense of intactness and worth, and phases where some members of the group claim superiority and even attempt to separate from the mainstream, "inferior," society are a normal stage in the reclamation of human dignity. We don't need to worry about oppressed groups with little to no power singing their own exclusive praises, so much as we need to worry about the oppressors, people who have actual privilege and who wield real power, who drive oppressed people to that degree of exaggeration in search of their own worth and a little space where they can feel safer.

Another popular feature was the graphics we produced. One of my life's proudest accomplishments was designing the original "I am not a puzzle/I am a person" graphic with a human-shaped puzzle piece inside a red circle crossed with a slash—the "not" symbol used on road signs. The original still hangs on my living room wall. Every time I see this design reused, probably by people who know only that it is a classic image of the autistic civil rights movement, I am reminded that I do have children who will carry on the values I imbued them with long after I am gone: that graphic is one of my progeny.

We encouraged autistic people to write for our library, where we posted essays that did anything but reaffirm that we were defective children whose job it was to obey. It was in the library where we most clearly spoke truths that had not been often said: that almost all autism treatment is based on false models of autism, that the institutions that have grown up around autism engage in violence, sometime subtle violence, and often overt violence, against autistic people in order to obtain compliance, that parents are often part of the problem—that bearing an autistic child makes no one, *ipso facto*, a saint, and that actual parents of autistic children are, if anything, more inclined to engage in abusive behavior toward autistic persons than the average parent, as illustrated by the many cases where parents of autistic children murder their own children. Of course, there are also many allies of ours among the parents of autistic children. The point is that no one automagically becomes an ally by virtue of merely existing—if you want to be an ally of any oppressed group, and you have privilege, you have to choose to be an ally, educate yourself, and work at it. Just thinking that you deserve to be counted among the good guys is *always* insufficient.

5 Autistics.Org and Finding Our Voices as an Activist Movement   75

The End of Autistics.Org?

I began this adventure in middle age at a time when I was beginning to have significant health problems, and aging is no friend of chronic health issues. Eventually I could no longer maintain the website. I put it first into an archived state, trying to preserve what was already there. Unfortunately over time even this has decayed, so that the domain autistics.org currently links to an empty directory and autistics.org lives on only at the wayback machine. I still have the files, and if I have the opportunity I will try to restore them. But I am looking for people who would like to continue what we started. I don't consider autistics.org dead, and I certainly don't consider the domain to belong to me personally (as you might imagine I've had many offers to buy the name from parent-led charities and commercial entities), but to the autistic community. Perhaps the domain autistics.org will rise again, under the leadership of a new generation.

During the years when autistics.org was actively maintained, we were anything but organized. Initially we tried to operate the website on shared hosting, but the volume of traffic got us shut down, more than once. So we moved the site to the cheapest dedicated server I could find. I tried (and sometimes failed) to manage the server, on the strength of having been an early adopter of Linux. Amelia (Mel) Baggs wrote prolifically for the website. Joelle Smith tried to make sense of the mess I sometimes made of Unix. Phil Schwarz was around from the beginning to nearly the end. Many others contributed, technically, financially, and through content. I apologize that I don't remember all of your names, and almost certainly there would not be space here to give everyone credit. Feel free to point to this article and tell others "I was there," because what you did changed the world.

We are tied in a single garment of destiny. Autistic activism is but one branch of an eternal great struggle, to set right what is wrong, to lift up those who have been pushed down, and to make space for joy in this world.

Reference

1. King M. L., Jr. (1963). *Letter from Birmingham jail.* Retrieved April 16, from https://wikilivres.org/wiki/Letter_from_Birmingham_Jail.

Open Access This chapter is licensed under the terms of the Creative Commons Attribution 4.0 International License (http://creativecommons.org/licenses/by/4.0/), which permits use, sharing, adaptation, distribution and reproduction in any medium or format, as long as you give appropriate credit to the original author(s) and the source, provide a link to the Creative Commons license and indicate if changes were made.

The images or other third party material in this chapter are included in the chapter's Creative Commons license, unless indicated otherwise in a credit line to the material. If material is not included in the chapter's Creative Commons license and your intended use is not permitted by statutory regulation or exceeds the permitted use, you will need to obtain permission directly from the copyright holder.

6

Losing

Mel Baggs

June 30, 1995.

Alone. In a bare room. For new people. They've left me here while they do my paperwork. First time in a mental institution.

For hours, I explore every inch. The unbreakable window looks out on an overgrown courtyard full of windows. There's a small circular hole with an oval impact crater. I wonder what happened. I find some graffiti on the bed, where nobody could see it without lying on the floor. It reads, "FOR THE ONES WHO DIDN'T MAKE IT."

Throughout the wait, I become aware I still exist. I haven't disappeared. I agreed to admission because I thought disappearing might be easier than suicide. It didn't work. I'm still here.

Crazy people are supposed to disappear. People disappear when they go in the front door of an institution. But I'm still here. I don't understand. I'm afraid. I often get things wrong. I know in my bones I'll never be

M. Baggs (✉)
Burlington, VT, USA

normal. I must be crazy. But crazy people in institutions vanish off the face of the earth. Everything in my whole life has told me this.

The whole world can't be wrong. Institutionalized crazy people disappear. I'm crazy, I'm in an institution, yet I still exist. Something has gone wrong. I must've screwed up if I'm still here. My innards twist with a mixture of worry, guilt, and frustration. Underneath, a bottomless, nameless dread. I'm still here, can think and feel, am alive, am whole, am suffering, am aware. Something somewhere has gone terribly wrong. I don't know what to do.

The most disturbing thing I've found in this room isn't the starkness, the unbreakable glass, the hole, the graffiti, the weight of untold stories, or the location. It's myself. I'm still here.

June 30, 1995.
Saint John's Autism Listserv. Known in the autistic community as ADHL—Academic Dick Heads List. People debate whether some people with developmental disabilities belong in institutions. Cal Montgomery writes an email about why institutions are bad and nobody belongs there. Things move on. Cal believes he's lost the argument.

Fast forward to my early twenties. I debate people who think some of us belong in institutions. I'm different, they say. I'm a different kind of person from those people who slam their heads against walls.

I remember that first time in a mental institution. I discovered that slamming my head against a wall led to being tied to a table until my arms and legs went numb, then injected with a drug that immobilized me so much I couldn't open my mouth. They have no idea.

I read archives of the Saint John's autism list. I do that: I try to find the roots of things, things that happened before I was around. I find Cal's 1995 message:

> <redacted> wrote: (original message by <redacted> in italics, Cal's comments in normal font)
>
> *I worked for over 5 years at the Eunice Kennedy Shriver Center for Mental Retardation which is located at the Fernald State School just outside of Boston, Ma. (this is where Christmas in Purgatory was filmed). I worked...*

I have worked for nearly 5 years in various staffed apartments in the community, some of whose residents used to live at Fernald, Wrentham, or other state schools. I have been on the Fernald campus only once.

...under a contract to provide the medical care for all the residents at all the state schools in Mass. I became very familiar with all of the facilities, most of the residents, and married one of the staff. Like any...

I am intensely familiar with the people who live in the staffed apartments where I have or do work, especially since I generally work with people with several disabilities, or with severe or profound MR [mental retardation], or both. I know fewer people than you do, but probably far more intimately. I am not familiar with any of the large institutions, but I do know a wide variety of current and former staff both in the state institutions and in the community residences.

...large facility, it was not what it could have or should have been, but it had improved quantum leaps from when it was filmed. And no one has been admitted to them since the mid 70's. Most of the staff are kind, and...

That is true.

...caring individuals who really grew to love the people they cared for and treated each of them as individuals. Many of the residents had been placed...

I cannot say anything about "most" since I don't know enough people. But while some of the staff at institutions are kind, some are truly cruel. I know this from observing former State School staff who have obtained jobs in community-based programs, from hearing stories from former State School staff who were horrified at what they saw, and from hearing stories from former State School staff who fondly reminisced about beating up residents of the Schools, or dropping water balloons on people trying to navigate icy paths, or ... Some of the things that people have told me about almost proudly would make you sick.

This is not to say that you don't get some of these people in the community-based programs, as well. But I have worked for a number of the agencies around here and have done relief shifts in others, and I have yet to see the acceptance and approval of this sort of behavior that I hear existed in some sections of each of the Schools. In the staffed apartments, there is more monitoring—from families, housemates' families, and neighbors, if from no-one else—available.

I have worked with clients who had nightmares about the School they grew up in, who refused to visit with friends who still live in cottages on campus, because they will not go to the campus, and who will not even stay in the room with you if the word "Fernald" or "Wrentham" comes up. (This is clearly the more verbal end of my caseload.)

...there as children by families who had been told to forget that they had ever been born. This was the prevailing advice from the professionals in the 40's and 50's. They may not now have families who want to or who are able...

That is true.

...to care for them. The functional level of some of those individuals is...

That is true. But there are (or should be) other options. Other options can be found, if we make finding them a priority.

...breathing. The staff work very hard with them and rejoice when a resident...

I have never ever met a person with MR whose functional level was breathing. And, as I say, I have been working with people with severe or profound MR full time (sometimes two jobs simultaneously) for 5 years and I have worked with others dating back to when I was 12. I *have* met many people whose abuse or mistreatment or neglect was rationalized because "he's retarded and doesn't understand" or "his level of functioning is so poor that there's nothing we can do for him." I have worked with people who, according to charts, don't communicate, but who in fact get a great deal across, consistently (same methods, same message received), to people who know them. I have *not* met anyone who is able to provide medical care for as many people as you say you have who has had the time to get to know those people well enough to accurately assess their functional level, so I'm assuming you are relying on other people. Given the facts that I've seen a lot of data fabrication from staff who used to work at State Schools, that I've seen wildly inaccurate assessments in charts of people who have come out of State Schools, and that the staff at State Schools may have been working in a 2:20 or lower staff:resident ratio in any case, I tend to doubt *any* assessment of functionality that comes out of those places. They are often simply wrong.

…they have cared for and worked with learns to sip from a cup or smiles. The staff do their best to make it as good a place as they can.

I am certain that many, and possibly most, of the staff, do. That does not mean that they are capable of making it a good place. Most people do not want to go live in a nursing home, no matter how hard the staff try to make it a good place to be.

Remember 15 years ago when the state psychiatric hospitals were closed in many states. The persons responsible for the closings did it with the best of intentions but the closings resulted in problems and mistreatment of the former residents that were much worse than any treatment they could have ever received in the facility.

For some of the former patients of State Hospitals, this is true. For others, freedom from the Hospitals freed them to get their lives together (yes, I do know such people). In any case, it was very very poorly planned and executed. There is no denying that. But that does not mean that such a move has to be poorly planned and executed.

This is my fear of what may result from the closing of these other facilities. Will proper care be taken to insure that the tragedies will not happen again? Before closing the facilities are appropriate…

Well, that is the responsibility of those of us in the community, isn't it? That is the responsibility of you, and me, and my neighbors (who are clearly not going to take on that responsibility—there is a great deal of hatred for people with disabilities around here).

…residences and treatment in place for all of who are those being uprooted?

It isn't enough to just keep saying, we can't close the Schools because we don't have appropriate places to put people. We need to look at the fact that direct care staff in Massachusetts have not received Cost of Living pay adjustments in about seven years, and that some could make more money on Welfare [sic] than they can at their jobs. So capable, competent, or even trainable people are not being attracted to these jobs. We need to look at the fact that Massachusetts does not mandate adequate human rights trainings so that these staff don't even know what they are supposed not to do. We need to make licensing a process that reflects more than how well the paperwork is kept. QUEST [Quality Enhancement Licensure and Certification Survey Process and Tool, a survey and certification process for providers of services for people with developmental disabilities] is a start. Recent changes in the rules for medication administration are a start. We need to make appropriate residences a priority instead of an excuse.

Remember, for most of these individuals this is the only home they have ever known. And while being a resident in a state facility may not be the best life, it is better than the streets.

Just because they've never known a better home, or a home with more actual choices, does not mean they don't deserve to know one. And just because there are things that we (as a society) could do that are worse, does not

mean that we do not have an obligation to do better.

Just my opinion, but an opinion based upon an intimate knowledge of the state facilities in Massachusetts.

As mine is based on knowledge of group homes.

Cal

I'm floored. The same day I entered that place, Cal said people like me don't belong there. Everything connected. Someone was on my side, even though I didn't know. The arguments I'm having now were going on before I knew the community existed. I was exactly the person being discussed:

an autistic person in an institution while people were arguing whether we belonged in institutions.

I order my records. They come in a giant box. I'm shocked. I'm not always aware how people see me. The papers have words I never knew people said about me: Low functioning. Severe and complex developmental disability. Unsalvageable. Violence. Headbanging. No future.

A group of autism parents attack an autistic woman named Michelle Dawson. They say she's not autistic. They target her viciously. I'm part of the response.

I describe myself in third person, using the most pessimistic terms professionals used on me. I reveal I'm that person, that I'd far rather have Michelle Dawson, Cal Montgomery, Laura Tisoncik, Joelle Smith, or Larry Arnold, speaking on my behalf than these parents. I call it "Past, Present, and Future."

The Autism Society of America puts out a website called "Getting the Word Out." It uses black-and-white stock photos and tragic language to describe us. They say our existence destroys our families. They're "spreading awareness."

I write a parody based on "Past, Present, and Future" called "Getting the Truth Out." I take black-and-white selfies, mash together every negative description professionals have used on me, then reveal who I am, what I think, what issues we face, and why "Getting the Word Out" harms us. I emphasize how you can make anyone look terrible if you describe them selectively.

I ask others to contribute their photos, selectively negative descriptions, and actual opinions [1]. But I'm the only one. Instead of a website starting with multiple pictures you can click, it ends up just me.

"Getting the Truth Out" becomes better-known than I expect. I remain unaware of the overall effect this website has had.

What I remember is people:

- Not grasping it was a parody, especially after the ASA took "Getting the Word Out" down.
- Thinking I meant it as straight autobiography.
- Not noticing I was highlighting the dangers of using a selective pathological description of anyone.

- Thinking the selective descriptions and stereotypes must be true, or I'm claiming they're true.

But the message was what I got from Cal: Those of us making the online arguments, and those of us described pathologically, can be the same people.

I'm not a different person than the scared child who found she still existed despite being in an institution. Who slammed her head against the wall of that room until brutally restrained. Who attracted pathological descriptions that stripped her of her humanity and future. I didn't become a different person once I started saying, "Institutions are bad."

Every person who could be described as an autistic child banging their head on the wall of an institution, is a hell of a lot more than that. It conjures a stereotype, a story. But it's not a type of person. It's more how others see us than who we are.

I'm not good at knowing the impact of my writing and videos. Just as Cal thought he lost the argument in 1995, that's how I assumed things went with "Getting the Truth Out." I thought people hadn't understood what I was trying to say, that I'd failed to convey my message.

From what people tell me that's not true. Yeah, lots of people misunderstood. But lots of people got the message. They affected others, who affected others. Everything we do has ripple effects we may never know. Maybe the day I put up that website, some child described in similar words was in an institution. That child may discover my writing and make the same connections I made when I found Cal's writing. You never fully know the impact of your own actions. Some of it may stretch beyond your death.

In writing this, I've done my usual: Described specific situations, in detail. But I didn't write this to tell my story. Each thing I mention can apply to many people and situations that aren't identical to the one I describe. That's how all my writing works: I write the specific but aim for broad applicability.

Our actions matter. We may never recognize the full impact we have. Everything we do can have a profound effect on others. Remember that when you think you've lost the argument, that you've failed, that nobody understood. Most of the effects of our actions are unseen but important. Without Cal's single email, there'd have been no "Past, Present, and Future," "Getting the Truth Out," or "In My Language" [2].

I owe debts to different disability movements: Developmental disability (DD) self-advocacy, Deaf culture, psych survivors, the independent living movement, others. They have shaped everything I've ever done in the autistic community, far more than the autistic community itself has.

Like everyone I know who's been called a leader within the autistic community, I took in many perspectives from outside the community, and functioned within a broader sphere than one community. We're part of the disability rights movement. Other disability communities influenced everything I'm known for within the autistic community. Some things people say as "about autism," were not only never "about autism," but came out of things like the DD self-advocacy movement. "In My Language" was an act of DD solidarity with a girl with cerebral palsy whose parents mutilated her, not a statement about autism. I told CNN this. They edited it out, replaced with something I never said [3].

Whatever perspective we come from, we need to be prepared to think we've failed, to never know our full impact. Cal may've lost an argument in 1995, but he showed me a way of seeing myself and other disabled people that proved central in nearly everything I ever did that had an impact. I'm sure even my least pleasant contributions have had important effects I'll never know about.

Sometimes we think we've lost, but it's only the beginning of things we can't imagine. We have to do the right thing even when it looks like we're failing. One email can spark important things we'll never see from people we've never met.

References

1. Baggs, M. (2006, August 7) *Want more contributors to getting the truth out* (Web log post). Retrieved from https://ballastexistenz.wordpress.com/2006/08/07/want-more-contributors-to-getting-the-truth-out/.

2. Baggs, M. (silentmiaow). (2007, January 14). *In my language* (Video file). Retrieved from https://www.youtube.com/watch?v=JnylM1hI2jc&t=31s.
3. Gajilan, A. C. (2007, February 22). Living with autism in a world made for others. *CNN*. Retrieved from https://edition.cnn.com/.

Open Access This chapter is licensed under the terms of the Creative Commons Attribution 4.0 International License (http://creativecommons.org/licenses/by/4.0/), which permits use, sharing, adaptation, distribution and reproduction in any medium or format, as long as you give appropriate credit to the original author(s) and the source, provide a link to the Creative Commons license and indicate if changes were made.

The images or other third party material in this chapter are included in the chapter's Creative Commons license, unless indicated otherwise in a credit line to the material. If material is not included in the chapter's Creative Commons license and your intended use is not permitted by statutory regulation or exceeds the permitted use, you will need to obtain permission directly from the copyright holder.

Part II
Getting Heard

7

Neurodiversity.Com: A Decade of Advocacy

Kathleen Seidel

Introduction

I was unschooled in autism before 1999. When Oliver Sacks's article, "An Anthropologist on Mars," appeared in *The New Yorker* in 1993 [1], my husband, Dave, and I had thought of our youngest child—a precocious, eccentric, artistic, excitable puzzle-lover with a penchant for solitude. "Sounds familiar!" we said to each other, but saw no reason for special concern.

I read nothing more about autism until 1999, after our first meeting with his fourth-grade teacher.[1] Her classroom observations led us back to the Sacks article, which resonated anew. A psychologist encouraged us to obtain an evaluation, and after several months of testing, in mid-2000 our child was diagnosed with Asperger syndrome.

---

[1] Since mid-2006, I have consistently used masculine pronouns to refer to the person identified in many of my earlier advocacy letters as my youngest daughter.

K. Seidel (✉)
Peterborough, NH, USA

© The Author(s) 2020
S. K. Kapp (ed.), *Autistic Community and the Neurodiversity Movement*,
https://doi.org/10.1007/978-981-13-8437-0_7

I set about learning everything I could about autism in order to best understand his needs and articulate those needs to his teachers. I read works by Lorna Wing [2], Uta Frith [3], Tony Attwood [4], and Simon Baron-Cohen [5]. I attended AANE (Asperger/Autism Network) seminars. I participated in discussions on the St. John's Autism List [6]. I trawled the internet for information and created a database to store my bookmarks and notes. I am a librarian by training, and had already developed several websites; an autism portal seemed like a potentially useful and rewarding creative project.

That took a few years to happen. After losing my job in the Dot-Com Bust of 2000, I started a used bookselling business, and underwent a profound shift in my understanding of autism.

Engaging Neurodiversity

I first encountered "neurodiversity" in an article in *The Atlantic* describing online spaces created by autistic adults [7]. I loved the compassionate, inclusive flavor of the word, and its broad call for respect for people like my child. I tucked it into my memory, whence it emerged one evening in January 2001, as Dave and I were brainstorming ideas for domain names. As it turned out, "neurodiversity" was available, so we registered it on the spot.

Although bookselling temporarily derailed my autism website plans, I continued to read and squirrel away URLs, increasingly gravitating to work by autistic authors. I delved into essays by Jim Sinclair [8], Frank Klein [9], Larry Arnold [10], and Joelle (then Joel) Smith [11]. I discovered the writings of Michelle Dawson, who opposed efforts in Canada to mandate one form of behavioral training as "medically necessary" for autistic children [12]. Laura Tisoncik's and Mel (then Amanda) Baggs's "Institute for the Study of the Neurologically Typical" [13] (see Tisoncik, Chapter 5) made a strong impression, and both eventually became friends. Janet Norman-Bain's "Oops...Wrong Planet! Syndrome" website led to hours of exploration [14]. I gave Gunilla Gerland's *Finding Out about Asperger Syndrome* [15] to my child to help him understand his diagnosis. Every reading sparked new shocks of recognition.

I began to notice autistic traits in every interaction I had with my father, and my mother—a music and special education teacher–increasingly shared that recognition. A chemical engineer born in 1933, Dad contracted polio at the age of ten, and used crutches, braces, and a wheelchair thereafter. In our household, "disability" and mobility accommodations were ordinary, as was a certain interpersonal *je ne sais quoi*. Dad was an analytical thinker with minimal tolerance for his own frustration or others' emotionality; he was practical, brusque, and disinclined to social niceties. (One especially memorable phone call began in mid-sentence, as if I were mid-conversation with him a continent away.) I gradually recognized that I, too, lay somewhere along the banks of the "broader autistic phenotype": intense focus; bluntness; anxiety and occasional sensory overload; fondness for collecting, organizing, and diving deeply into subjects that interest me. Why had nearly a decade passed before Dave and I figured out that our youngest child was on the spectrum? Perhaps because he was not so different from his grandfather, or from me.

Neurodiversity.Com

The appearance of "neurodiversity" in the *New York Times* [16] signaled that the time had come to transform my database into a website. In May 2004, after a week of nonstop HTML writing, Neurodiversity.com was born. Over 100 different pages included "Positive Perspectives on Autism," "Girls & Women on the Spectrum," "The Question of Cure," and "Neurotypical Issues"; the site also featured a game, "Unmasking the Face" [17], based on Paul Ekman's book on nonverbal communication [18]. Later, I added a collection of papers on the history of autism research, including early works by Kanner, Asperger, and Lovaas [19].

From my original mission statement:

> My goal is to increase goodwill and compassion in the world, and to help reduce suffering. I seek to help reduce the suffering of autistic children and adults, who often face extraordinary challenges in many domains of life, challenges made more difficult by others' unrealistic expectations and demands, negative judgments, harassment and marginalization. I seek to

help reduce the suffering of family and community members who are bewildered and distressed by actions of and interactions with autistic people, and who are concerned for their own and others' safety and well-being. I seek to help increase the capability of educators and service providers to provide effective, respectful support for those on the autistic spectrum. My means of achieving that goal is to share some of the information that has helped us to move from a place of grief and stress to a place of recognition, understanding, and positive regard. [20]

Engaging Advocacy

Shortly after Neurodiversity.com launched, I noticed a visit from the Yahoo group AutAdvo; soon I was avidly participating in discussions with autistic men and women whose writings I had previously encountered. I enjoyed exchanges with Camille Clark, co-creator of the Autistic Adults Picture Project [21], who blogged from 2005 to 2007 as "Autism Diva" [22]; Jane Meyerding, author of "Thoughts on Finding Myself Differently Brained" [23]; Phil Schwarz, father of an autistic son and vice-president of AANE [24]; Kassianne Sibley (now Kassiane Asasumasu); Patricia Clark (d. 2005 [25]); Alyric (d. 2009 [26]); and many other members.

Through AutAdvo, I befriended Gayle Kirkpatrick, whose autistic son had been banned from their town's only playground, and traveled to Maine in August 2004 to attend hearings in the family's suit against the school district. Shortly before trial, I summarized many themes that would inform my subsequent advocacy work in "The Autistic Distinction":

> ... Those who value compassion must work to change the content and tenor of public discussion about cognitive difference. ... We must consider the negative impact on autistic citizens of the popular practice of referring to autism as an "epidemic," a "tragedy," a "plague," a "devastating scourge," a "catastrophe," or a "demon;" of the use of military metaphors such as "killing," "attacking" or "defeating" autism, and description of autistic children as "prisoners of war;" of comparison of autism with degenerative diseases such as cancer and diabetes; of the use of verbs such as "fester" to describe the autistic pattern of human development.

… Many parents experience the unveiling of autism as a grievous revelation. However, I am convinced that the negative impact on families and on autistic persons is increased and perpetuated by crisis-oriented descriptions of autism that focus on abnormality and deficit, that automatically characterize early education as a heroic "intervention" if the children being educated are autistic, that raise the specter of institutionalization simply because that is the way society has tended to address cognitive difference in the past, and that describe autism as something that must be destroyed.

Assertions that autism can and must be "cured" create unrealistic expectations, promote the exploitation of parents made desperate by dire predictions, and perpetuate a climate of negative judgment towards children and adults who are not or do not strive to become "indistinguishable from their peers," those who look and behave like the autistic people that they are.

… I seek a reconceptualization of cognitive difference, to the end that those who bear now-stigmatizing labels of "deviance," "disorder" and "syndrome," may live and manifest their individuality, distinctive interests, gifts and capacities with integrity, in a manner that comes naturally to them, free of pressure to become people they are not, free of the automatic assignation of inferior status; and that they may enjoy the respect of their fellow citizens, rather than disdain and exclusion.

Retrospective consideration of the lives of exceptional human beings offers credible evidence that the autistic distinction has persisted throughout history, and has been a valuable element of human culture. Genetic research indicates that at least twenty different genes can signal a predisposition to autistic development; autism is pervasively embedded in the deep structure of humanity. Psychological research indicates that autistic characteristics constitute an identifiable pattern of traits that are present in varying degrees throughout our entire species.

Autism is as much a part of humanity as is the capacity to dream [27].

Through AutAdvo, I also befriended University of Wisconsin-Madison psychology professor and fellow autism mom Morton Ann Gernsbacher. When news broke that University of Kentucky chemist Boyd Haley had dubbed autism "mad child disease," we created a "Petition to Defend the

Dignity of Autistic Citizens" that garnered hundreds of signatures after it was published on Neurodiversity.com [28]. I soon learned that Haley was a hero to many convinced that autism was vaccine-induced, and a proposed expert witness in legal proceedings consolidating thousands of claims filed by parents of autistic children. The experience strengthened my concern about the proliferation of anti-vaccinationist sentiment, litigiousness, and chronic outrage among parents of autistic children; the use of sensationalistic language to describe autism and autistic people; and the hostility toward autistic adults expressed by many proponents of an "autism epidemic."

I wrote to the Congressional Autism Caucus [29], and many more letters to editors, and documented all of the correspondence on the site. These were noticed; Neurodiversity.com was mentioned in the December 2004 *New York Times* piece, "How About Not Curing Us, Some Autistics Are Pleading" [30]. Later, I had encouraging exchanges with the head of the National Institute of Mental Health, who had stated that autism "robs a family of [a child's] personhood" [31], and the head of the University of California, Davis MIND Institute, whose fundraisers had broadcast insupportable claims of an "autism epidemic" [32].

After *Rolling Stone* published "Deadly Immunity," alleging a cover-up of evidence linking autism with vaccines, I wrote to its author, Robert F. Kennedy Jr., noting that not all parents regarded their children's autism as a consequence of wrongdoing [33]. As the publicity campaign escalated for David Kirby's book *Evidence of Harm* (EoH) [34], I wrote to question Kirby about the purpose of the discussion list created in connection with it, documenting many instances in which autistic children were disparaged and dissenting parents vilified [35]. I also corresponded with Lenny Schafer—proprietor of the EoH list and publisher of the *Schafer Autism Report*, where autistic advocates were denounced as "imposters who trivialize the catastrophic nature of real autism" [36, 37]. I did not expect that my letters to Kennedy, Kirby, and Schafer would provoke attitudinal change in their recipients, but persevered and published in the hope that visitors to Neurodiversity.com might reconsider their assumptions about autism, recognize the toxic nature of the crusade to equate autism with contamination, and increase their respect for autistic citizens of all ages.

Neurodiversity Weblog

In the summer of 2005, I established Neurodiversity Weblog to streamline publication and facilitate discussion of letters and articles on the site. During the blog's first year [38–42], I published writing by Darold Treffert, Rita Jordan, Michelle Dawson, Phil Schwarz, and James Laidler. I deconstructed a *Chronicle of Higher Education* article condoning discrimination against autistic candidates for academic employment [43, 44]. I protested the Autism Society of America's pathos-inducing "Getting the Word Out about Autism" campaign [45], the use of vaccine-causation evangelists as consultants by Autism Speaks [46], and their doom-laden film *Autism Every Day* [47]. The blog attracted both sympathetic and critical readers, and featured many lively and occasionally contentious exchanges.

A February 2006 article in the *Concord Monitor* [48] spurred me to investigate a new regimen involving administration of hormonal suppressants such as Lupron to autistic children [49]. I learned that applications to patent the "Lupron protocol" remained undisclosed in the presentations of its developers, Mark and David Geier [50]; that the latter's academic affiliation was misrepresented in a peer-reviewed study [51]; that they headed the IRB overseeing their own research [52]; that they were diagnosing autistic children with precocious puberty who did not meet formal criteria [53]; and that the "protocol" called for excessive blood draws and expensive tests [54]. After publishing a series of articles on the subject [55], I sent the editorial board of *Autoimmunity Reviews* a lengthy critique of a report they had published of the Geiers's research [56]. This led to the paper's retraction, an incident later discussed in the *British Medical Journal* [57] and *Slate* [58].

During this period, I wrote to the Interagency Autism Coordinating Committee, calling out their lack of autistic members and their references to autism as a disease, and contrasting the scant attention they paid to quality of life issues with their frequent appeals for brain tissue, which gave the impression that autistic adults were more highly valued when dead than alive [47]. I also exchanged correspondence with the United Methodist Church protesting their support of *Sykes v. Bayer*, litigation initiated by a Methodist minister against corporations she held at fault for causing her son's autism [59].

In the fall of 2007, a paralegal studies course inspired me to write a series of articles on cases alleging environmental causation of autism. *Vaccine Court Chronicles* attracted heat [60]. On March 24, 2008, I published a post discussing economic incentives that biased vaccine-injury attorneys' pronouncements on disability causation, and tallied fees paid by the court to Geier associate Clifford Shoemaker [61]. Barely four hours later, Shoemaker issued a subpoena for me to be deposed in *Sykes v. Bayer*, demanding my financial records, tax returns, information about my religious beliefs, and all correspondence about any subject discussed on Neurodiversity.com [62].

After I blogged my motion to invalidate the subpoena, all hell broke loose. *Slashdot* covered the case three times, nearly crashing the server with each flood of hits [63–65]. *Opinionistas* across the blogosphere offered their support. Attorney Paul Levy of Public Citizen submitted a brief recommending that Shoemaker be sanctioned [66]. Harvard's Digital Media Law Project featured the case on their site [67]. I was profiled by the *Concord Monitor* [68], and in Andrew Solomon's *New York* article, "The Autism Rights Movement" [69], Dr. Paul Offit devoted a chapter to my work in *Autism's False Prophets*, and included me in its dedication [70]. After my motion was granted [71], *Sykes v. Bayer* was dismissed with prejudice [72], and the court ultimately sanctioned Mr. Shoemaker [73].

Over the next few years, I continued to report on ongoing autism-vaccine litigation, post announcements of research participation opportunities, and investigate dodgy autism treatments and consumer scams. Subjects included OSR, an industrial chelator developed by Boyd Haley and promoted for consumption by autistic children [74], and electromagnetic radiation shielding devices touted as autism treatments by debt-ridden multilevel marketers and new-age entrepreneurs [75]. A misleading telephone solicitation provoked me to dig into the public filings of the Autism Spectrum Disorder Foundation, which claimed to help autistic people, but showed little evidence of useful activity [76]. My local paper published an op-ed in which I advised readers to be skeptical of unfounded claims about autism [77].

In 2010, my investigations of the "Lupron protocol" and OSR inspired (and were cited in) the *Chicago Tribune*'s award-winning series on unproven autism treatments [78, 79]. Shortly thereafter, the FDA ordered

OSR taken off the market [80].[2] In the spring of 2011, following a Maryland citizen's complaint incorporating my articles on the "Lupron protocol," the state's Board of Physicians suspended Mark Geier's license to practice medicine and charged David Geier with practicing medicine unlawfully [82]. By 2013, Dr. Geier's license had been revoked in all twelve states that had granted it [83]. As the first person to raise the alarm about the Geier's pharmaceutical experimentation on autistic children, and about Haley's efforts to bypass federal drug approval regulations, I take pride in these outcomes.

Engaging Community

Neurodiversity.com was, for the most part, a one-woman operation; the occasional conference enabled me to connect with others who shared my interests and perspective. I was grateful to AANE for offering support to children like my son and parents like me. I learned much from AutCom's workshops on assistive communication and from accounts of its members. At Autreat 2008, I met Rosalind Picard of the MIT Media Lab [84], which I later visited with Mel Baggs and Estée Klar, founder of The Autism Acceptance Project [85]. I attended AutCom 2007 and Autreat 2009 as Mel's support person, traveling with them and assisting at their presentations.

As the amount of information and misinformation about autism proliferated online, updating Neurodiversity.com's static link pages grew increasingly laborious, and ended in 2008. The flow of blog posts lessened thanks to newfound employment; my son's labor-intensive adolescence and gender reassignment; my parents' passing; and advocacy burnout exacerbated by often-hostile attention attracted by my writing, and by the escalation of conflict between autism advocates.

In late 2005, Kevin Leitch, proprietor of the blog Left Brain/Right Brain [86] had established Autism Hub, a feed aggregator "guided by the principles of ethics, empowerment, advocacy, autism acceptance, positivity, and realism" [86]. The Hub eventually included a few dozen sites and served

---

[2]Since the FDA's 2010 order, OSR appears to have made a comeback [81].

as a useful portal to positive autism advocacy. In 2007, Kev needed to cut back on his obligations, so Dave assumed responsibility for Hub maintenance. A mailing list for status updates soon morphed into a discussion forum. Over the next few years much constructive conversation ensued; increasingly, so did "horizontally hostile" exchanges. Some list members questioned the legitimacy of others' diagnoses. Several protested the continued inclusion of a blogger publishing misogynistic posts; their protest provoked misogynistic responses. Some members asked to refrain from verbal abuse griped about supposed intolerance of their "autistic communication style." Resentment was expressed that a neurotypical parent was maintaining the Hub and moderating the list. Ari Ne'eman repeatedly challenged Michelle Dawson regarding Autism Speaks's funding of Laurent Mottron's research on autistic cognition, in which she was involved [87]. I was dismayed by the emergence of this last conflict, given my respect for Dawson's work, although I agreed that Autism Speaks should retool its goals and rhetoric.

I felt that needless discord, demagoguery, and polarization could only exacerbate tensions and undermine advocacy efforts, but I had too much on my plate to jump into new debates. Dave and I were both increasingly burdened by these conflicts, as well as by a member's depressive crisis. Stressors beyond the list included efforts to discredit Mel Baggs, whose video "In My Language" [88] had attracted overwhelming media attention. In May 2010, after one too many bouts of *agita*, Dave took the site offline, leaving it to others to proceed without his involvement. Many were unhappy about the abrupt shutdown, but sometimes one's own sanity must come first, and we do not regret our decision.

I continued to blog as I have described above, and attended one more conference. I published my last post in March 2012 [89]; one year later, a botched server migration vaporized Neurodiversity Weblog. Fortunately, the posts can still be accessed via the Internet Archive (http://www.archive.org). Although most of Neurodiversity.com's external links are defunct, I continue to host the site as a document of autism advocacy, the debate over autism and vaccines, and the evolving idea of "neurodiversity."

Although I own the web domain, I am reluctant to define "neurodiversity," preferring to express in writing the values I associate with it. The thoughtless deployment of stigmatizing characterizations of autism; the misleading marketing of unproven "autism treatments" to parents of autistic children; the litigation and culture of blame into which so many families were drawn—all presented morally and intellectually compelling matters for consideration, inspired by my conviction that cognitively variant persons of all ages should be afforded respect, appropriate assistance, and freedom from abuse, exploitation, and undue pathologization of their traits and challenges.

Conclusion

My path to advocacy began with the need to understand my child, and to marshal understanding within the school and community. I found the greatest insight for this work in writings of and interactions with autistic adults and their allies, both in person and online. With Neurodiversity.com and Neurodiversity Weblog, I sought first to share useful information, then to communicate my evolving concerns and encourage consideration of the concerns of autistic people themselves. I did not seek to join a movement, but ended up participating in one. As I put it in a 2006 letter to the *New York Times Book Review*:

> The neurodiversity movement does not consist of faddish cultists trolling for converts, but of disabled individuals, their family members, and [other] allies constructively responding to prejudice, stigma and pejorative labeling. People don't all think the same way, and appreciation of this reality is not limited to those who possess a diagnosis. I am one parent who wants her "neurodiverse" family members to flourish — happily, healthily, well-educated and respected in a society that embraces the value of cognitive variety. [90]

References

1. Sacks, O. (1993, December 27). An anthropologist on Mars. *The New Yorker, 69*(44), 106–125. Retrieved from https://www.newyorker.com/magazine/1993/12/27/anthropologist-mars.
2. Wing, L. (1998). *Autistic children: A guide for parents.* New York, NY: Kensington Publishing Corp.
3. Frith, U. (2003). *Autism: Explaining the enigma.* Malden, MA: Blackwell Publishing.
4. Attwood, T. (1997). *Asperger syndrome: A guide for parents and professionals.* London, UK: Jessica Kingsley Publishers.
5. Baron-Cohen, S., & Cosmides, L. (1997). *Mindblindness: An essay on autism and theory of mind.* Cambridge, MA: MIT Press.
6. St. John's University. (1999–2006). *Autism and developmental disabilities list* (Archived website). Retrieved from https://web.archive.org/web/20060113082637/http://maelstrom.stjohns.edu/archives/autism.html.
7. Blume, H. (1998, September). Neurodiversity: On the neurological underpinnings of geekdom. *The Atlantic.* Retrieved from https://www.theatlantic.com/magazine/archive/1998/09/neurodiversity/305909.
8. Sinclair, J. (1993). Don't mourn for us. *Our Voice, 1*(3). Retrieved from http://www.autreat.com/dont_mourn.html.
9. Klein, F. (2003, March 13). *Autistic advocacy* (Archived website). Retrieved from https://web.archive.org/web/20040319200741/http://home.att.net:80/-ascaris1.
10. Arnold, L. (1999–2001). *Kingdom of Laurentius Rex* (Archived website). Retrieved from https://web.archive.org/web/20040612094600/http://www.geocities.com/CapitolHill/7138/index.htm.
11. Smith, J. (2001–2006). *This way of life.* Retrieved from http://www.oocities.org/growingjoel.
12. Dawson, M. (2004, January 18). The misbehaviour of behaviourists: Ethical challenges to the autism-ABA industry. *No autistics allowed.* Retrieved from http://www.sentex.net/-nexus23/naa_aba.html.
13. Tisoncik, L., & Baggs, M. (1999–2004). *Institute for the Study of the Neurologically Typical* (Archived website). Retrieved from https://web.archive.org/web/20040611073537/http://www.autistics.org/isnt.
14. Norman-Bain, J. (1995–2005). *Ooops....wrong planet! syndrome* (Archived website). Retrieved from https://web.archive.org/web/20040506210446/http://www.isn.net:80/-jypsy.

15. Gerland, G. (2000). *Finding out about Asperger syndrome, high-functioning autism, and PDD.* London, UK: Jessica Kingsley Publishers.
16. Harmon, A. (2004, May 9). Neurodiversity forever: The disability movement turns to brains. *The New York Times.* Retrieved from https://www.nytimes.com/2004/05/09/weekinreview/neurodiversity-forever-the-disability-movement-turns-to-brains.html.
17. Seidel, K. (2004). *Unmasking the face* (Computer game). *Neurodiversity.com.* Retrieved from http://neurodiversity.com/nvc/index.html.
18. Ekman, P., & Friesen, W. V. (1975). *Unmasking the face: A guide to recognizing emotions from facial expressions.* London, UK: Prentice-Hall.
19. Seidel, K. (Ed.) (2004–2005). *Library of the history of autism research, behaviorism & psychiatry* (Article index). *Neurodiversity.com.* Retrieved from http://neurodiversity.com/library_index.html.
20. Seidel, K. (2004b). About us (Webpage). *Neurodiversity.com.* Retrieved from http://neurodiversity.com/aboutus.html.
21. Norman-Bain, J., & Clark, C. (2004). *Autistic adults picture project* (Archived website). Retrieved from https://web.archive.org/web/20040607004517/http://www.isn.net/~jypsy/AuSpin/a2p2.htm.
22. Clark, C. (2005–2007). *Autism Diva* (Archived weblog). Retrieved from https://web.archive.org/web/20080515193436/http://www.autismdiva.blogspot.com.
23. Meyerding, J. (1998). Thoughts on finding myself differently brained (Archived webpage). *M. Jane Meyerding Home Page.* Retrieved from https://web.archive.org/web/20050404121729/http://mjane.zolaweb.com:80/diff.html.
24. Schwarz, P. (2000). Personal accounts of being a university student with HFA/AS (Webpage). Retrieved from http://www.users.dircon.co.uk/~cns/phil.html.
25. Seidel, K. (2005, July 17). Farewell, friend: Patricia E. Clark, 1944–2005 (Archived weblog post). *Neurodiversity Weblog.* Retrieved from https://web.archive.org/web/20130419015004/http://neurodiversity.com/weblog/article/6.
26. Alyric. (2005–2009). *A touch of alyricism* (Archived weblog). Retrieved from https://web.archive.org/web/20130424135836/http://alyric.blogspot.com.
27. Seidel, K. (2004, August 20). The autistic distinction. *Neurodiversity.com.* Retrieved from http://neurodiversity.com/autistic_distinction.html.
28. Seidel, K. (2004, October 9). Petition to defend the dignity of autistic citizens (Webpage). *Neurodiversity.com.* Retrieved from http://neurodiversity.com/mothers_for_dignity.html.

29. Seidel, K. (2004, September 22). Autism and human rights (Webpage). *Neurodiversity.com*. Retrieved from http://neurodiversity.com/congressional_autism_caucus_letter.html.
30. Harmon, A. (2004, December 20). How about not "curing" us, some autistics are pleading. *The New York Times*. Retrieved from https://www.nytimes.com/2004/12/20/health/how-about-not-curing-us-some-autistics-are-pleading.html.
31. Seidel, K. (2005, April 4). Autism and personhood (Webpage). *Neurodiversity.com*. Retrieved from http://neurodiversity.com/autism_and_personhood.html.
32. Seidel, K. (2005, March 28). The "autism epidemic" and real epidemics (Webpage). *Neurodiversity.com*. Retrieved from http://neurodiversity.com/mind_epidemic.html.
33. Kennedy, R., Jr. (2005, July 14). Deadly immunity. *Rolling Stone*. Retrieved from https://www.rollingstone.com/politics/politics-news/deadly-immunity-180037.
34. Kirby, D. (2005). *Evidence of harm: Mercury in vaccines and the autism epidemic*. New York, NY: St. Martin's Press.
35. Seidel, K. (2005, May 29). Evidence of venom (Webpage). *Neurodiversity.com*. Retrieved from http://neurodiversity.com/evidence_of_venom.html.
36. Schafer, L., et al. (2005, January 11). *Schafer Autism Report* (Archived electronic newsletter). Retrieved from https://web.archive.org/web/20050507231402/http://lists.envirolink.org:80/pipermail/sareport/Week-of-Mon-20050110/000350.html.
37. Seidel, K. (2005, January 15). Lenny Schafer's inquisition: A response (Webpage). *Neurodiversity.com*. Retrieved from http://neurodiversity.com/inquisition.html.
38. Dawson, M. (2005, August 1). A few questions from Michelle Dawson (Archived weblog post). *Neurodiversity Weblog*. Retrieved from https://web.archive.org/web/20130419015000/http://neurodiversity.com/weblog/article/25.
39. Jordan, R. (2005, August 21). Is autism a pathology? (Archived weblog post). *Neurodiversity Weblog*. Retrieved from https://web.archive.org/web/20130419015013/http://neurodiversity.com/weblog/article/45.
40. Laidler, J. (2005, July 27). Chelation and autism (Archived weblog post). *Neurodiversity Weblog*. Retrieved from https://web.archive.org/web/20130419004919/http://neurodiversity.com/weblog/article/14.

41. Schwarz, P. (2005, August 4). Autistic people speaking about autism (Archived weblog post). *Neurodiversity Weblog*. Retrieved from https://web.archive.org/web/20130419015203/http://neurodiversity.com/weblog/article/26.
42. Treffert, D. (2005, August 10). Autism: Indeed new in name only (Archived weblog post). *Neurodiversity Weblog*. https://web.archive.org/web/20130419013009/http://neurodiversity.com/weblog/article/31.
43. Brottman, M. (2005, September 16). Nutty professors. *The Chronicle of Higher Education*. Retrieved from https://www.chronicle.com/article/Nutty-Professors/2488.
44. Seidel, K. (2005, October 1). Autopsy of a violent diagnosis (Archived weblog post). *Neurodiversity Weblog*. Retrieved from https://web.archive.org/web/20130131101931/http://neurodiversity.com/weblog/archives/54/.
45. Seidel, K. (2005, September 19). Getting the truth out about autism (Archived weblog post). *Neurodiversity Weblog*. Retrieved from https://web.archive.org/web/20130419013006/http://neurodiversity.com/weblog/article/51.
46. Seidel, K. (2007, January 18). On "Unstrange Minds" and the "autism epidemic" (Archived weblog post). *Neurodiversity Weblog*. Retrieved from https://web.archive.org/web/20130419004721/http://neurodiversity.com/weblog/article/123.
47. Seidel, K. (2007, January 15). Response to the Interagency Autism Coordinating Committee (Archived weblog post). *Neurodiversity Weblog*. Retrieved from https://web.archive.org/web/20130419005320/http://neurodiversity.com/weblog/article/122.
48. Conaboy, C. (2006, February 12). Did mercury cause Drew's autism? (Archived newspaper article). *Concord Monitor*. Retrieved from https://web.archive.org/web/20060305223725/http://www.concordmonitor.com/apps/pbcs.dll/article?AID=/20060212/REPOSITORY/602120310/1022/LIVING02.
49. Seidel, K. (2006, February 19). Autism and Lupron: Playing with fire (Archived weblog post). *Neurodiversity Weblog*. Retrieved from https://web.archive.org/web/20130401040535/http://neurodiversity.com/weblog/article/83.
50. Seidel, K. (2006, April 3). Patent medicine (Archived weblog post). *Neurodiversity Weblog*. Retrieved from https://web.archive.org/web/20130401051843/http://neurodiversity.com/weblog/article/94.

51. Seidel, K. (2006, June 9). An inaccurate byline (Archived weblog post). *Neurodiversity Weblog*. Retrieved from https://web.archive.org/web/20130401051843/http://neurodiversity.com/weblog/article/97.
52. Seidel, K. (2006, June 20). An elusive institute (Archived weblog post). *Neurodiversity Weblog*. Retrieved from https://web.archive.org/web/20130401051843/http://neurodiversity.com/weblog/article/98.
53. Seidel, K. (2006, June 26). A dubious diagnosis (Archived weblog post). *Neurodiversity Weblog*. Retrieved from https://web.archive.org/web/20130401051843/http://neurodiversity.com/weblog/article/99.
54. Seidel, K. (2006, August 25). Blood and data (Archived weblog post). *Neurodiversity Weblog*. Retrieved from https://web.archive.org/web/20130401051843/http://neurodiversity.com/weblog/article/110.
55. Seidel, K. (2006, August 25). *Significant misrepresentations: Mark Geier, David Geier and the evolution of the Lupron protocol* (Archived series index). *Neurodiversity Weblog*. Retrieved from https://web.archive.org/web/20130401040535/http://neurodiversity.com/weblog/article/109.
56. Seidel, K. (2006, November 17). Letter to Autoimmunity Reviews (Archived weblog post). *Neurodiversity Weblog*. Retrieved from https://web.archive.org/web/20130419002523/http://neurodiversity.com/weblog/article/116.
57. Deer, B. (2007). What makes an expert? *British Medical Journal, 334*(7595), 666–667.
58. Allen, A. (2008, January 30). Can vaccines cause autism? *Slate*. Retrieved from https://slate.com/news-and-politics/2008/01/eli-stone-scares-parents-and-angers-doctors.html.
59. Seidel, K. (2007, April 12). A plaintiff in the pulpit (Archived weblog post). *Neurodiversity Weblog*. Retrieved from https://web.archive.org/web/20130419003329/http://neurodiversity.com/weblog/article/126.
60. Seidel, K. (2008, June 2). *Vaccine court chronicles* (Archived series index). *Neurodiversity Weblog*. Retrieved from https://web.archive.org/web/20130401040535/http://neurodiversity.com/weblog/article/162.
61. Seidel, K. (2008, March 24). The commerce in causation (Archived weblog post). *Neurodiversity Weblog*. Retrieved from https://web.archive.org/web/20130401040535/http://neurodiversity.com/weblog/article/149.
62. Seidel, K. (2008, April 3). Subpoenaed (Archived weblog post). *Neurodiversity Weblog*. Retrieved from https://web.archive.org/web/20130401040535/http://neurodiversity.com/weblog/article/150.
63. Slashdot. (2008, April 11). Blogger subpoenaed for criticizing trial lawyers (Newsgroup post). Retrieved from https://yro.slashdot.org/story/08/04/11/1812225/blogger-subpoenaed-for-criticizing-trial-lawyers.

64. Slashdot. (2008, April 22). Blogger successfully quashes subpoena (Newsgroup post). Retrieved from https://yro.slashdot.org/story/08/04/22/1939219/blogger-successfully-quashes-subpoena.
65. Slashdot. (2008, June 24). Lawyer who subpoenaed blogger Seidel sanctioned (Newsgroup post). Retrieved from https://news.slashdot.org/story/08/06/24/1232258/lawyer-who-subpoenaed-blogger-seidel-sanctioned.
66. Seidel, K. (2008, May 27). Pure hearts and empty heads (Archived weblog post). *Neurodiversity Weblog*. Retrieved from https://web.archive.org/web/20130419005306/http://neurodiversity.com/weblog/article/160.
67. Ardia, D. (2008, April 11). *Sykes v. Seidel* (Archived webpage). Digital Media Law Project, Berkman Center for Internet and Society. Retrieved from https://web.archive.org/web/20140330234822/http://www.dmlp.org/threats/sykes-v-seidel.
68. Sanger-Katz, M. (2008, April 27). A forceful voice in autism debate (Archived newspaper article). *Concord Monitor*. Retrieved from https://web.archive.org/web/20080501163032/http://www.concordmonitor.com/apps/pbcs.dll/article?AID=/20080427/FRONTPAGE/804270376.
69. Solomon, A. (2008, May 25). The autism rights movement. *New York*. Retrieved from http://nymag.com/news/features/47225.
70. Offit, P. (2008). *Autism's false prophets: Bad science, risky medicine, and the search for a cure*. New York, NY: Columbia University Press.
71. Seidel, K. (2008, April 21). Quashed! (Archived weblog post). *Neurodiversity Weblog*. Retrieved from https://web.archive.org/web/20130404143304/http://neurodiversity.com/weblog/article/152.
72. Seidel, K. (2008, May 2). Mercury fades: Sykes v. Bayer dismissed (Archived weblog post). *Neurodiversity Weblog*. Retrieved from https://web.archive.org/web/20130401040535/http://neurodiversity.com/weblog/article/155.
73. Seidel, K. (2008, June 23). Sanctioned (Archived weblog post). *Neurodiversity Weblog*. Retrieved from https://web.archive.org/web/20130401040535/http://neurodiversity.com/weblog/article/164.
74. Seidel, K. (2008, August 1). A fine white powder (Archived weblog post). *Neurodiversity Weblog*. Retrieved from https://web.archive.org/web/20130401040535/http://neurodiversity.com/weblog/article/168.
75. Seidel, K. (2009, January 12). Who wants to be a millionaire? (Archived weblog post). *Neurodiversity Weblog*. Retrieved from https://web.archive.org/web/20130401040535/http://neurodiversity.com/weblog/article/181.
76. Seidel, K. (2009, March 12). Dialing for autism dollars (Archived weblog post). *Neurodiversity Weblog*. Retrieved from https://web.archive.org/web/20130419002321/http://neurodiversity.com/weblog/article/187.

77. Seidel, K. (2010, April 15). On autism: A word of caution (Archived weblog post). *Neurodiversity Weblog*. Retrieved from https://web.archive.org/web/20130419005312/http://neurodiversity.com/weblog/article/204.
78. Tsouderos, T. (2009, May 21). Miracle drug called junk science. *Chicago Tribune*. Retrieved from https://www.chicagotribune.com/lifestyles/health/chi-autism-lupron-may21-story.html.
79. Tsouderos, T. (2010, January 17). OSR #1: Industrial chemical or autism treatment? *Chicago Tribune*. Retrieved from https://www.chicagotribune.com/lifestyles/health/chi-autism-chemicaljan17-story.html.
80. Seidel, K. (2010, July 22). OSR: Off the market (Archived weblog post). *Neurodiversity Weblog*. Retrieved from https://web.archive.org/web/20130401040535/http://neurodiversity.com/weblog/article/209.
81. Carey, M. (2017, April 5). *Remember the fake supplement OSR #1? It's still being developed* (Weblog post). Retrieved from https://leftbrainrightbrain.co.uk/2017/04/05/remember-the-fake-supplement-osr-1-its-still-being-developed.
82. Mills, S., & Callahan, P. (2011, May 4). Md. autism doctor's license suspended. *The Baltimore Sun*. Retrieved from https://www.baltimoresun.com/health/bs-hs-autism-doctor-20110504-story.html.
83. Hawai'i Medical Board (2013, April 12). Board's final order. *In the matter of the license to practice medicine of Mark R. Geier, M.D. MED 2011-79-L*. Retrieved from http://web.dcca.hawaii.gov/OAHadmin/PDF_INDEX/OAHPDF/MEDICAL%20BOARD/MED-2011-79-L%20MARK%20R.%20GEIER%20M.D.PDF.
84. MIT Media Lab. (2008). *Affective computing*. Retrieved from https://www.media.mit.edu/groups/affective-computing/overview.
85. Klar, E. (2006). *The autism acceptance project* (Archived website). Retrieved from https://web.archive.org/web/20060623112134/http://www.taaproject.com:80.
86. Leitch, K. (2003–2007). *Left brain/right brain* (Archived website). Retrieved from https://web.archive.org/web/20071011015013/http://leftbrainrightbrain.co.uk.
87. McIlroy, A. (2011, November 2). The autistic advantage: Montreal team taps researchers' potential. *The Globe and Mail*. Retrieved from https://www.theglobeandmail.com/life/health-and-fitness/the-autistic-advantage-montreal-team-taps-researchers-potential/article4182520.
88. Baggs, M. (silentmiaow). (2007, January 14). *In my language* (Video file). *YouTube*. Retrieved from https://www.youtube.com/watch?v=JnylM1hI2jc&t=31s.

89. Seidel, K. (2012, April 18). Geier suspension upheld (Archived weblog post). *Neurodiversity Weblog*. Retrieved from https://web.archive.org/web/20130401040535/http://neurodiversity.com/weblog/article/221.
90. Seidel, K. (2006, February 15). One for the Times (Archived weblog post). *Neurodiversity Weblog*. Retrieved from https://web.archive.org/web/20130401040535/http://neurodiversity.com/weblog/article/78.
91. *Autism Hub* (Archived website). Retrieved from https://web.archive.org/web/20070102230042/http://www.autism-hub.co.uk.

Open Access This chapter is licensed under the terms of the Creative Commons Attribution 4.0 International License (http://creativecommons.org/licenses/by/4.0/), which permits use, sharing, adaptation, distribution and reproduction in any medium or format, as long as you give appropriate credit to the original author(s) and the source, provide a link to the Creative Commons license and indicate if changes were made.

The images or other third party material in this chapter are included in the chapter's Creative Commons license, unless indicated otherwise in a credit line to the material. If material is not included in the chapter's Creative Commons license and your intended use is not permitted by statutory regulation or exceeds the permitted use, you will need to obtain permission directly from the copyright holder.

8

Autscape

Karen Leneh Buckle

In the Beginning, There Was a List

I am currently the event manager for Autscape, an annual three-day residential event for autistic people. It is autistic-led, though not exclusive. I was there the day Autscape was conceived, and I have had a major role in organizing the event for 12 of its 14 years. Here's how it happened.

Soon after I first discovered the Internet in late 1996, I sought out autism-related groups and immediately went to spending hours every day on autism chat rooms on Internet Relay Chat (IRC) and email-based support groups ("lists") such as Independent Living on the Autistic Spectrum (InLv; Martijn Dekker [1], see Chapter 2) and Autism [2]. Through these activities, I heard about Autreat and joined ANI-L, the list run by the organization responsible for Autreat, Autism Network International (ANI, www.autismnetworkinternational.org). In those days, it was common to use an online "handle" or nickname rather than one's real name,

---

K. L. Buckle (✉)
Chesire, UK
e-mail: kalen@worldapart.org

© The Author(s) 2020
S. K. Kapp (ed.), *Autistic Community and the Neurodiversity Movement*,
https://doi.org/10.1007/978-981-13-8437-0_8

and mine was 'Kalen.' Then in its second year, Autreat was an annual conference organized by and for autistics and their "allies." I would have liked to go. At that time I had only met two people I knew were autistic, and I wanted to see if others were similar. I also wanted to see for myself what people were talking about so much on ANI-L. Unfortunately, as I lived in western Canada, and had very little money, the trip to the eastern United States was too daunting and expensive. When I emigrated to the UK, any hope of making it to Autreat all but disappeared.

One of the email groups I belonged to in those days was Independent Living on the Autistic Spectrum (InLv), run by Martijn Dekker. On InLv, we occasionally talked about Autreat, as some members had attended while others wished to, and we longed for someone to organize such an event in Europe. Eventually, in July 2004, tired of all the talk and no action, one member asked who would start a list (email group) to get started organizing such an event. I had some relevant experience running email lists for autistic parents, partners, and adults, so I volunteered. Considering myself radically disorganized, my intention was only to manage the technical side of the communication while others got on with the organizing.

Initially, the list had around 20 subscribers from all over the world, and it took some time to settle into a focused group of organizers. I was both flattered and astonished when a Finnish member, Heta Pukki, suggested that I should be chair. No one else came forward, so eventually I agreed. As a somewhat strange side effect of all of this coming from the online community is that my online name became my name within Autscape, so there I am still known mostly as "Kalen" except to my family and friends. Charles Burns was our first Treasurer, and Heta served as the first secretary. Various other people also helped with website, printing, and sharing thoughts and ideas, but I have such a poor memory for people I don't know who all of them were now. Jim Sinclair, the founder and main organizer of Autreat, served as an advisor to the committee. We chose a name and I created a website. We devised a list of features we needed in a venue, and Charles and his wife, Kazumi, found one that seemed viable, Ammerdown Centre in Somerset, England. They negotiated a discount, got us penciled in for 3 days in the summer of 2005, and put forward the initial deposit. Kazumi served as our first venue coordinator, managing liaison with the venue. However, interest waned and progress slowed, and we started to consider having a smaller event just for the organizers to learn about organizing events, with a view to doing a full-fledged conference the following year.

A pivotal moment came when early in 2005 Heta announced that she had secured a grant of £5000 to pay organizers' expenses and childcare, as several of us had children, and to bring over the leader from Autreat, Jim Sinclair, to assist and train us. Inspired by this change, we suddenly went into full swing organizing the actual event. I managed most of the general coordination, registration, program, and writing. I sent out a call for proposals for the program by email and made an advertising brochure. I also developed a rudimentary online booking form that would allow participants to send us essential information by email. Remarkably, people registered! Charles managed payments and banking. We didn't yet have a proper database for managing participant details, so I kept them all in a spreadsheet on my computer.

Autistic Space

Planning an event around autistic needs is complex due to their diversity. Because the event included room and board, Kazumi liaised with the venue staff at Ammerdown to obtain extensive details of bedrooms, bedding, meeting rooms, lighting, and menus. Although we had catering for special diets, we had a small number who couldn't cope with centralized catering, so we also had to arrange self-catering for them. We had to train venue staff in understanding autistic needs. We visited and took pictures of meeting rooms, social space, and bedrooms. To help those with sensory hypersensitivities, we banned scented products, camera flashes, and any touch without explicit consent. We scoured the venue for flickery fluorescent lights, squeaky doors, air fresheners, and noisy fans, and, wherever possible, had these switched off, fixed, or removed. We also advised participants to use sunglasses, earplugs, or other devices to suit them. Ammerdown had a lounge area with a bar where we could socialize in the evening. It had a similar feeling to a small pub, but without the background music because dealing with a room full of people talking is hard enough without music as well.

We adopted and adapted Autreat's interaction badge system, large colored badges optionally worn by participants to help regulate social approaches. They were:

Red: Please do not approach me. I do not wish to socialize with anyone.
Yellow: Please do not approach unless I have already told you that you may approach me while I am wearing a yellow badge.
Green: I would like to socialize, but I have difficulty initiating. Please feel free to approach me [3].

On each of the badges was written the name of the color, to help those with color blindness, and its use, in case participants forgot what it was for.

We decided from the start that the event should be three days long so participants would have a chance to settle in and still have some time left to enjoy being there. Coffee and meal breaks were extra long for similar reasons. Because many autistic people appreciate structure, there was a full program of scheduled activities such as presentations and discussions. The program also included a leisure time after lunch when a variety of activities were on offer, such as music, dance, and meditation, and we had an exercise session early in the morning for those who wanted to join in. We knew that many of our participants would not drive and public transport to the venue was poor, so we arranged transport by bus from the nearest train station, 13 miles away. Because autistic people have a wide variety of gender identities and expressions, we now gender neutralize toilets in public areas when there isn't already a gender-free one, but we didn't know to do that at the time.

One thing that makes a space autistic is that autistic people are ordinary, not special. Autscape is a prime target for researchers and journalists, and while some autistic people are eager to take part, others wish to be left to be themselves in peace. We have had some journalists, filmmakers, and researchers on site, but we give them strict guidelines disallowing filming in public places and to only use material from people who proactively offer to contribute.

Having put time and effort into organizing all this for the first event, we then hit a snag. The venue decided we would have to provide one carer for every three autistic people. Most of our potential participants were independent adults who didn't have or need someone to care for them. From this came one of our first principles: participants are presumed competent. If they need help to manage in Autscape's environment, it is up to them to bring someone, although as it turns out, people who usually need a carer when away from home sometimes don't need one at Autscape. Thankfully, an emergency meeting with venue staff was sufficient to convince them that we didn't need to dictate a specific carer-to-participant ratio. Autreat organizers had warned us that some difficult behavior from participants was likely and Jim provided some training just before the event.

Possibly the most helpful adaptation for autistic people is the provision of information. Before Autscape, we sent out a comprehensive yet concise information pack with details of the venue and program, expectations, packing list, and transport details. We were, as always, late producing it, only 3 weeks before the event, but one participant has said that after some considerable anxiety, it was the information pack with its detailed information that told her she was coming to a place where she would be understood.

Most of the bedrooms had two single beds, but demand was so high we needed maximum occupancy, so we arranged to match up roommates and assign rooms ourselves based on sensory and social preferences. We filled all of our designated bed spaces in the venue, which was shared that year, for a total of around 45 people. The first event was a stunning success. Participants seemed happy, organizers were coping, the program ran well. There was no sign of the difficult behavior we'd been told to expect. We had a meeting at the end to consider whether Autscape had a future (and who would run it if it did) and the response was overwhelmingly positive.

Autscape continued in much the same way for the next two events, but with sole use of the venue so we could fit in about 85 people if we used every space. In two of the years, we even had people camping in the grounds in order to come. By 2008 we had outgrown Ammerdown, and it had become too expensive as they no longer gave us the discount for new groups. We searched through hundreds of retreat centers from web searches and recommendations, but all were unsuitable. We eventually settled on Giggleswick School in North Yorkshire, England. As we were told to expect based on Autreat's experience, numbers dropped by about a quarter when we changed venues. Giggleswick is on a steep hill and bedrooms and facilities are scattered among several buildings. We didn't know how to get a coherent sense of community in a dispersed venue, so to many who had been to Autscape before, it felt more distant and detached. In 2009, we returned to the retreat center format at Emmaus Centre near London, and then back to Ammerdown for the last time in 2010. After the moves, numbers increased back to Ammerdown's full capacity. Since then, we have become better at choosing venues and managing their different strengths and weaknesses, so different needs are met each year and we are better able to provide social opportunities even in dispersed venues.

Autistic Participants

Inclusivity is a central principle of Autscape. In order to be inclusive, Autscape has to be accessible. The main criticisms Autscape has received over the years have been from two views of accessibility. One is that it is insufficiently accessible to people with mobility impairments, and the other is that it is too expensive for autistic people, who very often have a low income. Unfortunately, we have not yet been able to solve both of these at once. Venues with full wheelchair access, adapted bedrooms, en-suite bathrooms, etc., are expensive. Fees would have to be considerably increased to cover such venues. Autscape has to pay the same fee to the venue whether the person is a participant or a personal assistant, so allowing assistants to attend for free would also substantially increase fees. Inexpensive venues are usually boarding schools that are hundreds of years old. They tend to be more difficult to access, work with, and navigate. They have a less polished appearance and provide fewer luxuries, so participants have to bring their own towels and use communal kitchenettes for making coffee and tea. Rather than being compact like most conference centers, schools are often on sprawling sites with long walks between activities, dining areas, and houses of residence. Within the houses, they have many stairs, convoluted corridors, awkward height beds, and few or no en-suite bathrooms. Although the external door to each house is locked with a separate code, the bedroom doors usually don't lock. However, these venues usually have great outdoor space, lots of separate rooms to meet in, and far more single bedrooms, which many of our participants need. Our best solutions to these problems for now are to keep our program and advertising costs to a minimum and to rotate between venues with different features to prioritize different needs in consecutive years.

Even in the cheaper venues, the total cost for Autscape is still high, but such a high proportion of our participants have a low income that lowering the fees for them would put an unfair burden on the few who don't. Until 2017, we offered a very small low-income discount. The discount had to be small in order to cover a large number of people. We never required proof of income to avoid disadvantaging those who struggle with organization and paperwork. This discount was not helpful to those who had the least and struggled to afford to come even at the lower rate. In 2018 we tried a new approach. We stopped having that low-income rate and instead have a grant scheme which is more flexible and allows us to give larger discounts to a few people rather than a very small discount to many. It is more expensive, but as a more mature organization, our finances are in better condition than they were in the early days.

Some of the issues with access could be solved by diligent fundraising in order to lower all the fees or provide more substantial subsidies for people who can't afford to come. Unfortunately, people with the ability to fundraise effectively appear to be rather rare in our community. We had three large grants in the early years. The first, as described earlier, funded organizers and childcare at the first event. After the second grant we had the misfortune of being randomly selected for an audit of the spending. The stress of the audit was enough to put us off seeking further grants. Funding for ongoing running costs of an organization is virtually impossible to find, but we did apply for one more grant in 2008 to support Autscape's development as an organization. Since then, Autscape has been entirely self-sustaining, with no external funding other than the occasional personal donation. Fees are set at a level that can sustain operation, and prudent management of the finances has allowed Autscape to remain well above the minimum necessary to operate since the first event.

Another aspect of being as widely inclusive as possible is Autscape's decision to be explicitly non-political. That includes party politics—which is not allowed under UK charity regulations in any case—and "autism politics" such as lobbying for changes to law or policy. The sole exception is that Autscape supports the view that autistic people have a right to exist. Autscape was set up exclusively to run the annual Autscape event. If we become distracted from that, we risk failing at our core purpose. We also risk losing some of our inclusivity. If we adopt a political position, then we alienate those who disagree with it. Autscape has provided a platform for more politically oriented activities to take place. Participants have run discussion groups about the Labour Party's Neurodiversity Manifesto and about autistic activism. Out of one of the latter, the London Autistic Rights Movement was started, which later evolved to be Autistic UK, an active autistic rights organization. One former member, who was involved in another organization for a while, has suggested that Autscape's focus and determination to do one thing and do it well, not to be distracted by other activities like politics and advocacy, is part of why Autscape has lasted so long.

Autscape has had minimal success in attracting autistic people with a wider range of ability. To some extent, this is due to inherent characteristics of people who are, for example, less verbal or less sociable. However, we have tried to include some activities that are more accessible to people who don't handle words as well as most of us, with mixed success. In 2013 I did a presentation on using Makaton, simplified sign language, to communicate urgent needs. We also try to include some sessions that are more experiential, such as meditation, art, music, and movement workshops. To do a really good job of having a program suitable for a wider range of autistic people, we would have to have a dedicated leader and team who focused on only that, but so far all our organizers have been quite busy just continuing to have an event each year.

Being inclusive does not mean making everyone happy. It is common for autistic people to believe that in a hypothetical autistic-only space (which Autscape isn't, but many people have talked about creating) everyone would be exceptionally sensitive and respectful of their needs. In our experience at Autscape, many autistic needs are mutually incompatible, even paradoxical. Some of us are loud, but easily overloaded by noise.

Some are lonely, but only want interaction on our own terms. Some can't stay still and quiet in presentations, but are intolerably distracted by others fidgeting or interrupting. It isn't possible to satisfy all of these at once.

In wider society, many autistic people are assumed, by virtue of their diagnosis, to be at risk of antisocial behavior. Social groups set up by neurotypicals (NTs) for autistic people sometimes require all their autistic members to sign a behavior contract before they are allowed to attend. Such contracts are seldom seen outside of activities for children, teenagers, and disabled people, and it is demeaning. Autscape decided early on that we would not have any kind of general behavior contract. Guidelines are framed in terms of challenges that we can work together to manage. Outright rules are kept to a minimum, mainly around venue house rules and limiting sensory distress to others. We don't require anyone to wear an identification badge, speak in a certain way, or refrain from acting odd or "creepy." We don't tolerate harassment, but when an issue arises, we are more likely to work on ways for both parties to understand each other and to avoid a repeat incident rather than coming down hard on the perceived transgressor. Although we try to limit the negative impact of others' behavior, autistic space is an exercise in tolerance.

Autscape has never excluded non-autistic people. Exclusivity supports the idea that such segregation is needed, that is, that the presence of people from outside our group would be somehow harmful, not "safe." Having inclusion criteria, for example, "autistics only" creates suspicion about whether those in the group are really "us" or may be "them," whether deliberately (infiltrators) or by mistake (falsely identifying as autistic). Exclusivity also lends itself to the spread of prejudice and misinformation about the excluded group, as statements about them can go unchecked, and allows those in the group to foster ideas about their superiority to the excluded one. It is inconsistent to respond to exclusion from mainstream society by practicing it. Some argue that exclusion is justified when the selected group is a disadvantaged minority, but we believe that for all the same reasons that neurodiversity is beneficial to society, it is beneficial to our group.

Herding Cats (Autistic Organizers)

Events for autistic people that are organized by NTs can be autistic-friendly, but they will never be truly autistic space. For better or worse, a great majority of Autscape organizers are autistic. One of the pleasures of working with an autistic team is that we have some communication compatibilities, such as focusing on content rather than the feelings behind it, and being swayed by a well-supported argument. Many Autscape participants don't think they are capable of being organizers, but many organizers didn't either. We have minimum standards of communication and decision-making abilities in order to be able to operate at all, but difficulties, even quite severe ones, are tolerated and accommodated as much as possible, and mistakes, even big ones, tend to be forgiven and eventually forgotten. Some who thought we couldn't work with anyone have done quite well in Autscape. For some of us, it is the only significant contribution we have been able to make to society.

As chair of Autscape, when it wasn't going well, I felt like a cat herder. Autistic organizers have a tendency to get bogged down in details, or go off on a preferred focus, neglecting the priority task at hand. Executive functioning difficulties are very common, and a few of us have severe initiation impairments as well, so we take on tasks and then aren't able to complete them on time (or at all). It is also distressingly common for committee members to fall completely silent when we are struggling. When it was going well, rather than a cat herder, I felt like an orchestra conductor, bringing out the unique features and capabilities of each member. One needs a lot of handholding, another needs an assignment and the freedom to get on with it. Some need constant reminders, others will grind to a halt in response to "nagging." I had to carefully balance demands, responsibility and support to bring out the best in each member.

In 2007, Autscape nearly ended. An interpersonal conflict escalated into a full division of the committee into two factions. It was brutal, messy, personal, and public. Those who believe that autistic people are always honest and loyal, and that groupthink, backbiting, and interpersonal politics are NT failings we are immune to, are wrong. We may not engage in these things with NTs on their terms, but in our own way, we are quite capable of all kinds of bad behavior. By the time the dust settled,

6 months after the conflict started, about half of the organizers had left Autscape permanently. They said that Autscape was fundamentally flawed and doomed to fail.

One of the keys to Autscape's long term success has probably been that it has been a democratic organization from the start. Although the conflict in 2007 led to the loss of many important and hardworking committee members, the organization did not rely wholly on any individual for its survival. That crisis was an essential turning point in Autscape's development that ultimately led to it becoming more mature and robust as an organization. The organizers who remained made some substantial changes to our ways of working that have persisted.

In order to improve transparency and avoid any future allegations of conspiracy or covert bullying, we formalized considerably, possibly excessively. Our informal methods of decision-making and on-list discussion gave way to monthly committee meetings in an online text-based chat, with agendas and minutes published on the website. Meetings have served as a good impetus for those who fail to do things without a deadline, and provide a forum to ask questions of those organizers who don't respond to email.

Our company secretary at the time, Yo Dunn, worked hard with a lawyer and a small group of volunteers to put together our new governing documents and register Autscape as a limited company and a charity. This process was finally completed in May 2011. The company registration was to limit the organizers' liability in case something went wrong. Prior to that, if Autscape had failed to happen for any reason, the individuals on the committee would have been responsible for any financial obligations to the venue. In 2015, we came up against another administrative barrier. If a charity has a turnover of more than £25,000 per year, they need to have an independent examination of the accounts. Our accounts were fine, but our documentation was not up to that, so we spent a year trying to keep our income down while we worked on getting the necessary procedures in place. This threshold has now been crossed, allowing us to choose larger and more expensive venues. It is also now possible to have more than the one event in a year, although that has yet to happen.

Autscape is autistic-*led*, but not exclusively autistic-*run*. We have, from the very start, usually had one or two people on the organizing committee who do not identify as autistic, but we have maintained an autistic majority without having to make rules or quotas to ensure it. Such a quota would require all board members to openly identify as autistic or not. Most board members already do so, but for various reasons, some people don't, and it is not Autscape's habit to force individuals to do anything. We don't fear an NT takeover, because the whole culture is based on autistic ways of working. In fact, we have had only minimal success at retaining NT members. Our non-autistic organizers are seldom actually neurotypical— they are always family members of autistic people who tend to have other neurodivergent conditions such as ADHD, or significant autistic traits. Having some non-autistic organizers can help with sourcing some of the skills that are uncommon in autistic people. We find they are often better at communicating with NTs, so they have made good venue coordinators, form fillers, and phone call makers. Neurodiversity, which means having NTs in autistic space as much as it does autistics in NT space, is a benefit to the organization.

As organizers, we have had to learn that an autistic Utopia is impossible. We often get suggestions or complaints from participants about things we are already aware of, but due to our disabilities and limited resources, have been unable to act on. For example, it has been suggested that Autscape be longer, that we have more events, that we fundraise more, or that we publicize the event more widely. Aside from any difficulty with doing tasks, advertising may increase demand and we are struggling to keep up with interest as it is. Quite often we have been told that we ought to have web forums instead of or in addition to the email list we provide for social interaction of participants between Autscapes, but doing this properly would require more of our tech people than they have to give right now, when they have other priorities and demands on their limited capacity. One of the most common suggestions in response to our occasional shortage of resources to continually do more and better is, logically, to take on more organizers. It may be counter-intuitive, but so far our experience is that the social and communication demands of a larger team require more resources than what they free up by taking on work.

Where Next

Currently, the biggest challenge to Autscape is meeting the demand. Autscape has been steadily growing throughout its life, but the last two years have shown a sharp increase to over 200 participants. With more participants' fees contributing to the program, we can select more presentations from the increased number of proposals received, which, in turn, cater for a larger audience. The main downside of all this growth is that it excludes those who can't cope with such a large group.

One way forward is to continue allowing Autscape to grow, but a more manageable solution may be to have more events. We have taken a step in that direction by separating the governance of the Autscape organization from the planning of the Autscape event, which will allow separate groups of organizers to manage each event. Autscape could then support events of different sizes, styles, and locations. They could potentially even be in another European country. This has been predictably complicated and, also predictably, is taking more time than anyone thought it would or should. However, we are now more or less prepared for another event when an organizing team is ready to take it on.

I remember at the very first Autscape, now 14 years ago, walking through the venue filled with autistic people chatting, laughing, learning, and generally enjoying themselves, and thinking to myself, "I did this." I found it almost unbelievable, even as I saw it myself. It worked. I didn't start Autscape to be part of a movement or to be an activist. I also didn't do it to be part of a community or to meet other autistic people; I don't even like meeting new people. I do, however, like creating a little autistic space where we can be the ones who are normal, a place where autistic people can meet, learn, socialize, have a good time, and feel they are understood. I started Autscape mostly by accident, but I continued because I wanted to make something that would be a positive influence in autistic people's lives. It is.

References

1. Dekker, M. (2010). *Independent living on the autism spectrum* (Electronic mailing list). Retrieved from http://inlv.org/inlv-historic.html.
2. St. John's University. (1999–2006). *Autism and developmental disabilities list* (Archives). Retrieved from https://web.archive.org/web/20060113082637, http://maelstrom.stjohns.edu/archives/autism.html.
3. Buckle, K. L. (2005). *Autscape 2005 information pack.* Retrieved from http://www.autscape.org/2005/infopack.doc.

Open Access This chapter is licensed under the terms of the Creative Commons Attribution 4.0 International License (http://creativecommons.org/licenses/by/4.0/), which permits use, sharing, adaptation, distribution and reproduction in any medium or format, as long as you give appropriate credit to the original author(s) and the source, provide a link to the Creative Commons license and indicate if changes were made.

The images or other third party material in this chapter are included in the chapter's Creative Commons license, unless indicated otherwise in a credit line to the material. If material is not included in the chapter's Creative Commons license and your intended use is not permitted by statutory regulation or exceeds the permitted use, you will need to obtain permission directly from the copyright holder.

9

The Autistic Genocide Clock

Meg Evans

Alarm Bells

When I first thought about creating a Star Trek fanfiction website in the summer of 1999, I had no idea that the site would later become known for a ten-year countdown timer warning of a potential genocide in the making. That summer, everything in my life seemed to be going very well. I had just found a good job after staying home with my kids when they were little, and fun stories were all I had in mind for the Ventura33 website—so named because I got the idea for it while driving along California's Highway 33 in Ventura, known formally as the City of San Buenaventura (Fig. 9.1).

The topic of autism, along with society's views of it, was not on my radar at that point in time. Although I had seen the word used in reference to me as a child, I thought it simply had to do with early childhood language

M. Evans (✉)
Dayton, OH, USA
e-mail: Meg@megevans.com

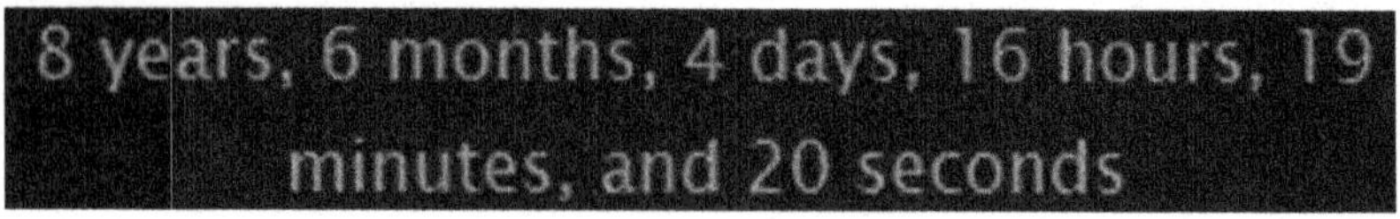

> 8 years, 6 months, 4 days, 16 hours, 19 minutes, and 20 seconds

Fig. 9.1 The Autistic Genocide Clock by Meg Evans

development. I knew that the fact I'd learned to read and talk at about the same time was unusual, but I didn't understand what relevance it might have in adult life. I was an early reader, while others in my extended family had been slow to speak, and I simply took it for granted that everyone developed at their own natural pace. People sometimes told me that my speech sounded a bit odd, which I attributed to living in different parts of the country as a child and getting my regional accents muddled. I didn't see that as significant either.

Sometime toward the end of 2002, I began to notice that there were sensational stories cropping up in the mainstream media about "Asperger syndrome," a now-outdated term that meant autism without a speech delay. I had not seen that term before and did not identify with it. The stories all followed the same general pattern of describing children who behaved in peculiar ways, thus supposedly causing their parents to lead lives of intolerable misery. At first, I paid very little attention to that narrative, dismissing it as a ridiculous pop-psychology fad that couldn't last long. After all, raising quirky children was certainly nothing new in the history of parenting. The children described in those stories didn't strike me as all that odd anyway.

Far down in my subconscious mind, though, a few dots started to connect. By late 2003, the picture had grown clear enough that my internal alarm bells were sounding. Those sensational articles hadn't gone away, but instead were showing up more often. Their scope was not limited to children but also encompassed autistic adults, who were commonly described as freakish, incapable, barely human, and unsupportable burdens on society. Internet searches only turned up more of the same, and I began to realize that I was looking at a dangerous mass hysteria.

Aspergia

One site that I came across during those searches was an exception—the UK-based forum Aspergia. Created in 2002, it aimed to spark critical discussion of society's attitudes toward autism by creatively framing the issue in terms of speculative fiction. Aspergia's featured story asked readers to consider: What if—rather than being defined by a medical label—autistic people were an ethnic minority group, descended from an ancient tribe with a recognized history and culture? Would society then be willing to accept, respect, and accommodate their differences? And if so, then why wasn't that happening in real life, and what needed to change?

Although the website's name obviously was derived from the term Asperger syndrome, the site was not exclusive to those who had received that particular diagnosis. The forum community welcomed all participants equally and sought to encourage a respectful conversation about what disability meant in relation to autism. Some members came to the site believing that autism was inherently disabling, while some did not view themselves as having a disability at all. Many informative discussions took place regarding the social model of disability, which holds that people become disabled not as the inevitable result of a physical or mental condition, but because socially constructed barriers prevent them from fully participating in society.

The existence of a forum where autism was discussed in terms of the social model of disability may not seem remarkable by today's standards; but at the time, many people had never seen anything like it. Some had grown up internalizing ugly stereotypes and myths, believing that they never would have a place in society. Although others had a vague sense that all was not as it should be, they couldn't quite say how. The conversation on Aspergia challenged participants to give more thought to the prevailing cultural assumptions. It was a daunting and often uncomfortable process of consciousness-raising.

Because I found it hard to understand why the culture was full of stories about autism that diverged so fundamentally from my own view of the world, I did some reading. I learned the awful history of what had been done to people with developmental disabilities in the twentieth century—eugenics, institutions, exclusion from schools and other public places.

Slowly it dawned on me that when I had changed schools several times as a six- and seven-year-old, the reason hadn't been—as I naively assumed at the time—to try out different schools and see which one was a good fit, like trying on clothes at the store. My mother, when I asked her, confirmed that the school administrators had told her I was not welcome to stay, and so she kept trying until she found a school that would keep me. This was a few years before the federal government required schools in the United States to educate all children.

Aspergia's forum closed in July 2004. The site was a casualty of its own success, in that it had grown much faster than its administrator—who called himself Edan—had anticipated. As a consequence, the site never had enough moderators to deal with the frequent arguments and flame wars that inevitably resulted from challenging people's worldviews.

Continuing the Conversation

Several former members of Aspergia, who believed that the conversation needed to continue, started building their own websites. The most successful in terms of sheer numbers was Wrong Planet, a forum site created by two teenagers, Alex Plank and Dan Grover, who envisioned a welcoming social space. Discussion of the more politically charged issues often took place in a forum in the UK called Aspies for Freedom (AFF), founded by Amy and Gareth Nelson. AFF was designed to include not only autistic activists but also parents concerned about their children's future.

On the AFF parenting forum, non-autistic parents sought advice to help them better understand their children's needs. Autistic members offered suggestions, while also discussing civil rights concerns in other areas of the site, where parents were welcome to participate if they wished. I took part in some of these conversations under my forum nickname of Bonnie Ventura. Describing myself as a person who belonged to a multigenerational autistic family, I gently encouraged parents to recognize their own autistic traits and to trust their instincts in raising their children. Among those who joined AFF in early 2005 was Kevin Leitch, a British parent who soon built the Autism Hub blog aggregator with the goal of constructively bringing together parent bloggers and autistic activist bloggers.

I'd had the Ventura33 story website for about five years by then, and it had grown into an archive that included not only my stories but also those of several other contributors. Inspired by Aspergia's use of fiction as a catalyst for discussion of autism in the context of disability rights—a concept that was becoming more commonly known as neurodiversity— I decided to create a page on Ventura33 for that purpose. I put out a call for stories featuring autistic characters and others with neurological differences in the Star Trek universe, which I posted not only on fanfiction writers' lists, but also on AFF and other autistic community forum sites.

My goal for the neurodiversity page was simply to encourage my readers to think a little farther outside their cultural boxes. I didn't anticipate that Ventura33 would play a major role in bringing together autistic activists to organize for civil rights in real life. Rather, I had in mind that the stories would promote reflection and constructive dialogue as a counterweight— if only a small one—to society's unthinking repetition of autism myths.

Countdown

Then I came across a disturbing news article published on February 23, 2005, which left me with a greater sense of urgency. Entitled "Autism research focuses on early intervention," it began by discussing studies of siblings' behavior and then moved on to government funding and genetics. The author interviewed Dr. Joseph Buxbaum, head of the Autism Genome Project at the Mount Sinai School of Medicine, and discussed his expectation of "major progress in identifying the genes associated with autism in the next decade" [1].

That in itself did not immediately strike me as cause for concern—after all, one might expect a research scientist to be optimistic about work in progress, and genetic research could potentially have many different aims. But then I scrolled down a little farther and found this unambiguously stated prediction:

> Buxbaum says there could be a prenatal test within 10 years.

What I found most unsettling about this statement was not simply the fact that it had been made, but that the worldview from which it sprang was devoid of meaningful examination. The overall tenor of the article—and, indeed, of the general public discourse surrounding autism at the time—was that everyone agreed the world should not have autistic people in it. The only question, as many saw it, was how to reach that goal. An entire layer of critical inquiry into the underlying assumptions had been effectively short-circuited.

I started composing a response to post on Ventura33, along with a link to the article. I wrote that the possibility of a prenatal test for autism raised significant ethical concerns. This was not an issue of abortion politics, as I saw it; rather, it had to do with informed decision-making and the value that our society places on different kinds of people. Government funding to develop a prenatal test, together with stereotypes and misinformation in the media that characterized autism as a devastating burden to families and society, gave rise to a coercive environment in which pregnant women would not be able to make truly informed decisions.

Because many autistic people go undiagnosed, I wrote, the total number worldwide was likely to be much higher than was generally believed—perhaps over 100 million. (More recent scientific estimates have confirmed that this higher number is in fact likely.) This would be equivalent to about one-third of the US population, or the total populations of the UK, Canada, and Australia combined. As such, prenatal testing for autism would amount to eugenics on the largest scale in human history.

Ending with a call to action, I asked my readers to visit other autistic advocacy websites and, if possible, to create their own; to get involved in real-life advocacy events; and to contact policymakers to express their views. Consistent with the ongoing dialogue in the AFF forum community, I urged parents to work toward building a society where their children's lives would be valued.

My working title was "Autism Research and Prenatal Testing." That title seemed too bland, though; it didn't convey a feeling of urgency. I asked my husband, who is a software developer, to add a timer at the top of the page counting down 10 years from the date of the news article. After he added the code, "The Autistic Genocide Clock" was launched on May 22, 2005.

Drawing Attention

I posted the first group of stories on the neurodiversity page in June 2005 and got some comments by email. One of them was a question from a student named Ari Ne'eman—was I involved in any real-life civil rights organizations focusing on autism? No, I was not, I answered; but in my opinion, such organizations were much needed.

Because Ventura33 was only a small fanfiction website, I wasn't expecting either the neurodiversity page or the clock page to get much attention. The site was so small, in fact, that my husband had put both it and his personal blog on a server in the basement using our basic residential Internet connection. The server was just an old, slow desktop computer that I had bought as surplus from my employer for 20 dollars, but that was good enough because we got so little traffic.

It took me a while to realize that my site had in fact drawn more attention than I'd anticipated. In late 2005 and early 2006, I did occasional Google searches on the word "neurodiversity," looking to see what new activist websites had emerged. I noticed that the Ventura33 neurodiversity page was consistently in the top ten results. At first, I assumed that was because the concept was still new enough that there hadn't been much written about it yet.

Then one day, my husband said, "Hey, Meg, did you know that so many people have been deep-linking to your clock page that we're running out of bandwidth?"

I told him, no, I hadn't been aware of that. My husband was keeping detailed statistics, though, and there was no doubt the clock page was getting most of the increase in traffic. Eventually, we ended up moving our websites to a virtual private server.

Autism Hub

Meanwhile, Autism Hub had gotten underway; there were about fifteen blogs in the aggregator in early 2006, and it grew rapidly from there. Several of its members had medical or other science backgrounds, and the early Hub posts often warned about the dangers of quack treatments purporting to cure autism. Disability rights topics were a large part of the discussion too. The Hub's initial members included Joelle (then Joel) Smith, whose list of autistic murder victims was a precursor to the Disability Day of Mourning.

The Hub's parent bloggers—one of whom was Estée Klar, a Canadian art curator who founded The Autism Acceptance Project and promoted inclusion by way of the arts—often wrote about happy moments in their everyday lives. At that time, images of autism in the mainstream media had been so relentlessly negative that even these simple, cheerful posts about enjoying family life went a long way toward changing the narrative for the better.

In May 2006, the Hub's bloggers erupted in outrage following the release of a video entitled *Autism Every Day* by Autism Speaks, which was then a newly formed organization. The video depicted the lives of families with autistic children as a fate literally worse than death; one parent featured in it said that she had thought about driving off a bridge with her autistic daughter in the car. The producer, Lauren Thierry, suggested that most parents of autistic children had such thoughts at one time or another. Autism Hub promptly created an online petition entitled "Don't Speak for Us," and many of the petition's signatories commented on the risk that the video and other similar depictions in the media might actually incite child murder.

The informal community of bloggers at Autism Hub had a significant impact in bringing disability rights issues surrounding autism into the public consciousness. Although well-funded organizations such as Autism Speaks largely dominated the discourse in 2006 and 2007 through traditional media, by this time society was becoming aware that other views existed. Autistic activists felt more empowered to assert themselves in the public sphere and to envision a future without the barriers created by ignorance.

Moving Toward Acceptance

The Autistic Self Advocacy Network (ASAN) was incorporated in November 2006 as a nonprofit organization run by and for autistics. Seeking to address public policy issues relating to autism from a disability rights perspective and to teach self-advocacy and leadership skills, it began as a small, all-volunteer group that declared "Nothing About Us Without Us." In December 2007, ASAN successfully organized an advocacy effort, together with Autism Hub's bloggers, which brought about the removal of billboards in New York entitled "Ransom Notes" that had depicted autism and other disabilities as evil kidnappers snatching children.

I was invited to join ASAN's board toward the end of 2008. During my first two years of service as an ASAN board member, it became apparent how quickly the culture was changing in the direction of acceptance. Pressure from autistic and cross-disability activists convinced Autism Speaks to disavow its September 2009 release of the video *I Am Autism*, which was another portrayal of autism as a child-snatching, family-destroying monster. The mainstream media and policymakers became more careful to use accurate and respectful language in referring to autism. President Obama appointed ASAN founder Ari Ne'eman to a term on the National Council on Disability that began in 2010.

I took down the original Autistic Genocide Clock page in July 2011 and posted a revised page [2] because I felt that autistic activism—including the efforts of Aspergia, AFF, Autism Hub, ASAN, and many others—had improved the culture enough so that routine prenatal testing for eugenics purposes would not be widely seen as desirable. Moreover, by then scientists had learned that the genetic factors involved in autism were very complex, which made it unlikely that any simple, routinely administered test would be developed. (Some tests do exist for single-gene conditions associated with autism, and this remains a concern to the extent they are used for prenatal testing rather than to confirm a clinical diagnosis.)

On the revised page, I wrote that although the cultural and political landscape had changed for the better in many ways since 2005, the history of that time period and the activists' determined efforts should not be forgotten.

References

1. Herera, S. (2005, February 23). Autism research focuses on early intervention. *CNBC*. Retrieved from http://www.nbcnews.com.
2. Evans, M. (2011, July 16). *Autism research and prenatal testing*. Retrieved from https://www.ventura33.com/clock.

Open Access This chapter is licensed under the terms of the Creative Commons Attribution 4.0 International License (http://creativecommons.org/licenses/by/4.0/), which permits use, sharing, adaptation, distribution and reproduction in any medium or format, as long as you give appropriate credit to the original author(s) and the source, provide a link to the Creative Commons license and indicate if changes were made.

The images or other third party material in this chapter are included in the chapter's Creative Commons license, unless indicated otherwise in a credit line to the material. If material is not included in the chapter's Creative Commons license and your intended use is not permitted by statutory regulation or exceeds the permitted use, you will need to obtain permission directly from the copyright holder.

10

Shifting the System: AASPIRE and the Loom of Science and Activism

Dora M. Raymaker

My Introduction to the Autistic Advocacy and Neurodiversity Movement

When I co-founded the Academic Autism Spectrum Partnership in Research and Education (AASPIRE) in the summer of 2006, I was already grounded in my identity as an Autistic person, autistic rights activist, and general troublemaker. That grounding had taken time, however, and struggle. Growing up, I had been the stupid one, the confused one, the damaged, scary, aloof, broken, worthless, lazy, crazy, alone, alone, alone, one-of-a-kind, never-trying-hard-enough, busted, always alone one—and I had made peace with that. I had learned to love it, and worn it for a skin. Then, in 1999, with the tsunami-like smack of a co-worker's incorrect assumption that I already knew what I was, I learned I was *none of those things*. I was Autistic. Even positive, healthy change—as this was—can

---

D. M. Raymaker (✉)
School of Social Work/Regional Research Institute for Human Services, Portland State University, Portland, OR, USA
e-mail: draymake@pdx.edu

be traumatic if it requires redefining Self. It is the disintegration of the old identity that is terrifying, not the truth of the new one. My journey of redefining Self started with my friends and co-workers presenting me with the truth, reached a climax with a clinical diagnosis during a period of intense crisis, and ended with my embrace of the neurodiversity movement.

When I first encountered the neurodiversity movement in the early oughts through the writings of Jim Sinclair, Mel Baggs, Joelle Smith, and others on the autistics.org [1] site and disability/autistic rights blogs and forums (like the This Way of Life [2] blog and the old LiveJournal Asperger's community group), I was not new to activism, nor to challenging the social order. I'd been radicalized (woke) in the gay (Lesbian, Gay, Bisexual, Transsexual, Queer, and others, LGBTQ+) rights movement of the late 1980s. I'd marched, created, subverted, and put my body at risk for my civil rights back then, starting with basics, like the right to go outside without being murdered. So when I first started exploring Autistic identity, the first thing that struck me was the similarity between being Autistic and being queer. Nothing brought me to the neurodiversity movement, and there was no choice involved: the world needs all minds just as it needs all genders, sexualities, and other vectors of diversity. As a systems scientist I know diversity provides social and ecological systems with flexibility, resilience, innovation, and a greater chance of optimal survival. As a human rights activist, I know all humans are to be valued. I always have, and always will fight for my communities of identity, and for other marginalized communities, because the empowerment of one benefits us all. I will always pay it forward from the activists who came before me to those who will come after; we are connected in a lineage of social justice.

Individuals and Organizations Critical to My Contribution to Neurodiversity and Autistic Activism

In the summer of 2006, I met AASPIRE's co-director, and my mentor and now long-time collaborator, Christina Nicolaidis—a physician-researcher and parent of an autistic child who also has experience in feminist and queer activist spaces. Christina, adorably, invited me to an autism scientific journal club because she wanted to meet me (apparently, my postings to the parent-focused Portland Aspergers Network e-mail list were helpful to her), and knew I would never say yes to unstructured social time. We read maybe three research articles, all of which provoked outrage, before deciding not to complain but to do something about the problem we were seeing. "The problem" being autism research that was poorly designed, stigmatizing, offensive, useless if not downright harmful, unethical, or otherwise failing to be of practical benefit to actual autistic people or the A/autistic community.[1] This is the very problem an approach to science called Community Based Participatory Research (CBPR) has been developed to solve.

CBPR is an approach to scientific inquiry that includes people from communities of identity as co-researchers in all phases of research that impacts their community, starting with deciding what to research in the first place [3]. The approach grew out of the field of public health in response to inequities experienced by communities defined largely by race or ethnicity [3]. CBPR has since been used with communities defined by many other identities; however, at the time of AASPIRE's founding, it had never been used with the A/autistic community. CBPR is an emancipatory approach to research, which explicitly acknowledges the connection between knowledge and power, and attempts to return power to communities that experience oppression. CBPR makes no attempt to decouple science and activism; instead, it seeks to use the rigor of science to disrupt the ways that science contributes, both directly and indirectly, to institutionalized oppression. Science can then become both better at answering questions about the world (i.e., be better science), and a vehicle of empowerment.

Coincidence of timing and circumstance is as much a factor in successful activist work as skill, and when Christina and I met at that autism journal club in my living room in the summer of 2006, she was using CBPR

---

[1] Capital "Autistic" is used to denote people who culturally identify as Autistic (but may or may not have an autism diagnosis) and lowercase "autistic" to denote people who may have a diagnosis but not culturally identify as Autistic. A/autistic is intended to be inclusive of both overlapping and interconnected communities of identity.

with communities of color to conduct culturally responsive depression care and interpersonal violence research. She'd had the same conversations about problems with traditional approaches to research with her African American and Latina partners that she was having with me about autism research. I was already an activist deeply committed to civil rights and grounded in the neurodiversity and disability rights paradigms. Thus, AASPIRE was born of a deep need for autism research that is driven by the A/autistic community, a recognition of the potential of science to empower the A/autistic community, and a once-in-a-lifetime collaboration between an Autistic activist (who grew up to be an academic researcher) and an academic researcher (with a history of activism) who ended up being best friends.

However, two individuals and a handful of friends do not make a research group, and so Christina and I went looking for additional collaborators, a process which took some time. CBPR typically forms partnerships between community-based service or policy organizations and academic institutions to co-conduct research, drawing on the strengths of their respective networks. The problem was there were so few Autistic-run organizations at the time, and those that were well-established, such as Autism Network International (ANI) and the Global and Regional Asperger Syndrome Partnership (GRASP), had missions focused elsewhere from service provisioning or policy change. The only community-based organization with a service or policy focus we knew of that was run by actual Autistic people was some brand-new thing called the Autistic Self Advocacy Network (ASAN) that we thought (in our own fledgling ignorance) may or may not even have been a real thing. Initial conversations seemed promising though, and by 2007 we had decided to collaborate. I met ASAN's co-founders Ari Ne'eman and Scott Robertson at the very first talk AASPIRE did at a scientific conference (the 2008 meeting of the American Association of Intellectual and Developmental Disabilities [AAIDD] in Washington, DC). At that time, AASPIRE had yet to complete a research study and didn't have any business, really, presenting at a research conference other than that we were the only people in the country doing community-engaged research with the A/autistic community. I remember sitting in my hotel with Ari and Scott as they asked me how I felt about the medical (deficit) model of disability. I fell out my chair—as

I am wont to do when not restraining my hyper-kinetic tendencies, hence my general mistrust of furniture—and proclaimed with the semi-echolalia I was relying on more heavily at the time, "Better dead than med!" They laughed, and I laughed, and we were all relieved that we were on the same neurodiversity page. They asked me to be on their board and ASAN was the first community-based organization to partner with AASPIRE; we've helped each other out ever since. In more recent years, AASPIRE has also developed a strong relationship with the Autism Society of Oregon; however, Autistic voice always has been, and always will be, privileged above organizations that are not self-advocate led. I am happy that there are now many more Autistic-led organizations serving a wide variety of missions.

Intended Goals of My Neurodiversity or Autistic Activism

Over the years, I've contributed to the neurodiversity movement and autistic activism in various ways, always with a goal of social justice. Some contributions have been successful: I yelled through a speech device at a town hall meeting, resulting in the governor adding a role for a self-advocate to the state's Autism Commission, and then serving in that role for two years. Others have been less successful: I blogged on autism issues for Change.org and mostly succeeded in getting death threats and paralyzing anxiety. The success of some contributions remain to be seen: I have several works of fiction published by Autonomous Press (autpress.com), a neurodivergent-run publisher, that center queer neurodivergent characters and attempt, as with pretty much all things I've ever done, to subvert systems of oppression [4, 5]. But out of all of the neurodiversity activism I've done (see dorararaymaker.com), my work with AASPIRE and in the sciences is both what I am most proud of, and what I feel has been the most successful.

My book chapter (written 2010) in *Worlds of Autism*, "Participatory research with autistic communities: Shifting the system" [6], details—in a wide-eyed, fresh-out-of-my-Master's-program way—the dynamic I still believe is at play between science, society, and community. Regardless of the struggles scientists may have in 2018's political climate, scientists still have more power than many people in oppressed communities. Because of the power society has to marginalize or center communities, empowering communities to influence science—and encouraging scientists to be influenced by communities—is, I believe, a point of leverage. In other words, if we think of behavior in a complex system (like human society) as generated by its structure, then changing the structure will change the behavior. Change the dynamic between scientists and autistic people, and the behavior of the entire system shifts.

Also, I'm a much better scientist than I ever was policymaker or politician, and far more suited to the calm of the lab than the drama of social media. There are plenty of peers in the movement who have those areas covered. What my scientific activism is intended to accomplish can be summarized in AASPIRE's mission statement [7]:

- To encourage the inclusion of people on the autism spectrum in matters which directly affect them.
- To include people on the autism spectrum as equal partners in research about the autism spectrum.
- To answer research questions that are considered relevant by the autistic community.
- To use research findings to effect positive change for people on the spectrum.

Steps to Meet the Intended Goals

AASPIRE conducts autism services research for adults. This is an area in which there continues to be a paucity of academic attention. The topics we focus on are an intersection of what the Autistic community prioritizes, and what we have the scientific expertise on our team to fund and successfully carry out. To date, this has primarily been in the field of health and mental health services, though recently we have branched out into employment, and are open to any type of research that meets our mission.

A lot has been written in the academic literature about how AASPIRE operates [8–10]. To summarize, we meet and communicate in ways that equalize power. We include a majority of autistic people on the team, and the whole team makes decisions about the research together. We respect each other's expertise and work hard at building trust and a safe—though not always comfortable—space for co-learning between autistic community members and academic allies. We have always had some members with intersecting identities in the group, for example, our co-directors include an autistic academic and a parent academic—which I think helps us translate respectfully between each other's cultures, and find effective ways to share power. We try to select new team members for people who prioritize getting the work done.

We also go through the steps typical of health and social services research: come up with an idea, obtain funding to realize it, carry out the research, disseminate the results to both the scientific community and the public (which, for us, includes the A/autistic communities). Part of the basic business of research also involves participation in professional meetings and conferences, engagement in academic and popular science forums like Reddit, and working with policy and scientific entities like the Interagency Autism Coordinating Committee (IACC) to shape research priorities and the direction of future funding.

It is through this normal business of science that I feel I, and AASPIRE, have had the most success in shifting the system. In a way, this book chapter is a "ten years after AASPIRE's founding" companion to my position in *Worlds of Autism* as it asks me to reflect: Have we shifted the system? Has engaging autistic people in autism research made any difference in the way society and public policy views the A/autistic community? Have we generated a behavioral change through our small influence on the broader structure of science? Have we made life any better for people in the A/autistic community?

Between 2006 and this writing in 2018, AASPIRE has obtained funding to complete a series of five healthcare studies, one employment study, start cutting-edge research on autistic burnout, and create an online, interactive Healthcare Toolkit for autistic adult patients, their supporters, and their healthcare providers (autismandhealth.org). Science moves slowly, and it can take a lot of time to see change from research and practice. However, our results indicate that both patients and healthcare providers found the Toolkit to have positive impact on their healthcare interactions [11], and we are continuing to find ways to implement the intervention effectively. In the community, numerous autistic people, parents, and clinicians have told us the Toolkit has made a positive difference in their lives. As we complete preliminary steps of learning how best to serve autistic people in employment and other aspects of well-being, we hope to continue to create interventions that directly help the community. Simply participating in CBPR can also help the community; an evaluation of a study which included many members of AASPIRE found that being a co-researcher is empowering and may enhance self-advocacy skills [12].

Members of AASPIRE have spoken on a variety of topics at the IACC, the US government body that sets federal priorities for autism research; people with close ties to, or then-collaborators with, AASPIRE successfully pushed that body to include autistic voice on their board—and sat on their board. I have witnessed multiple "a-ha" moments from scientists and students at my talks when they realize inclusive research is both possible and desirable. We have given our data to ASAN to use for making policy points. Both Christina and I have done highly attended Reddit Science Ask Me Anythings (AMAs) [13, 14] about AASPIRE's work, which has enabled us to reach the broader public with both neurodiversity ideas and the practical results of our research. We have been invited speakers at multiple conferences and events, either to discuss our findings or to discuss how to conduct inclusive research, more and more as time goes on. Our publications on neurodiversity have been loudly cited, while our publications on our research have quietly infused the next generation of academics interested in conducting research in collaboration with the A/autistic community (see aaspire.org for an ongoing list of AASPIRE's publications and activities). Recently, due in part to AASPIRE's visibility in the field, Christina was asked to start a new journal *Autism in Adulthood* (https://home.liebertpub.com/publications/autism-in-adulthood/646) that includes A/autistic people (both scientists and not, including myself as Associate Editor) in all aspects of its editorial processes. The power that I, and my colleagues, have as peer reviewers in other journals to reject articles that are disrespectful, stigmatizing, or poorly designed is significant. Strangers have told me how AASPIRE's work has impacted community—recently, an Autistic counselor at a medical training I did told me they had been using AASPIRE's Healthcare Toolkit for years with their clients. I have obtained my Ph.D. and become the first openly Autistic person to get funding from the National Institute of Mental Health (NIMH) to conduct autism services research; I recently returned from an NIMH meeting where I was able to discuss inclusive research with colleagues and policymakers. My existence is an act of resistance.

Ten years ago, any one of these things would have been unthinkable.

Ten years ago (or thereabouts), AASPIRE was receiving comments on its grant proposals like "there is not adequate evidence that the self-reports of individuals on the autism spectrum are valid or reliable" (Anonymous, 2010). Ten years ago, we in the autistic rights movement were fighting to get people to believe autistic adults existed at all. Now our work is the leading edge of a new movement of inclusive, participatory research with autistic people worldwide [15].

It is neither AASPIRE's nor my work alone that made change possible—it is the whole of the autistic, neurodiversity, and disability rights movement chipping away at the status quo, being relentless in its march toward social justice, and transmuting its collective rage into sacred anger to burn down everything in its way in order to build a more just and inclusive system. Nothing about us without us.

My Work's Place in the Broader Movement

AASPIRE works within the sphere of academic health and social services research to conduct projects the A/autistic community wants done. I have been capacitated by that work to extend more broadly into the wider neurodiversity and disability communities. I've worked with the developmental disabilities community on research examining connections between violence, disability, and health with the Partnering project [9, 16]. I've worked with women with intellectual disabilities and autistic women to understand their experiences with pregnancy and pregnancy decisions, and develop peer-led informational videos [17]. I've worked with the broader disability community on a peer-developed abuse awareness and prevention program [18]. I've worked with young adults who have experienced first episode psychosis on developing peer-created interactive online tools to reduce stigma and increase self-determination for others in their community [19]. I am the faculty advisor for my university's Disability Alliance student group, and they are doing important activist work at a local, university level [20]. Because, again, we are nothing if not a collection of intersections of our identities, and the empowerment of one benefits us all.

There is no wrap-up for this chapter, no succinct "lesson learned." AASPIRE, and my intersectional positioning as a neurodivergent, queer, and gender queer activist and scientist remains an active, ongoing experiment in whether or not restructuring the way science is conducted can shift the system toward justice for all.

What Neurodiversity Means to Me

Neurodiversity, to me, means both a fabulous celebration of all kinds of individual minds, and a serious, holistic acknowledgment of the necessity of diversity in order for society to survive, thrive, and innovate. It means identity, belonging, and community. It means I am not broken, not alone, and neither are my siblings standing with me beneath that huge, multi-colored neurodiversity umbrella: we the autistic, the mad, the weirdly-wired, the queer, the crippled, and the labeled with neurodivergent diagnoses like flowers that glorify our beautiful bodies and minds.

References

1. Autistics.org. (2011). *Autistics.org: The real voice of autism* (Archived webpage). Retrieved from https://web.archive.org/web/20140208025225/, http://archive.autistics.org.
2. Smith, J. (2001–2006). *This way of life* (Archived webpage). Retrieved from http://www.oocities.org/growingjoel/.
3. Israel, B. A., Eng, E., Schulz, A. J., & Parker, E. A. (2005). *Methods in community-based participatory research for health*. San Francisco, CA: Wiley.

4. Raymaker, D. M. (2018). Heat producing entities. In N. Walker & A. M. Reinhardt (Eds.), *Spoon Knife 3: Incursions* (pp. 215–248). Fort Worth, TX: Autonomous Press.
5. Raymaker, D. M. (2018). *Hoshi and the red city circuit.* Fort Worth, TX: Autonomous Press.
6. Raymaker, D. M., & Nicolaidis, C. (2013). Participatory research with autistic communities: Shifting the system. In J. Davidson & M. Orsini (Eds.), *Worlds of autism: Across the spectrum of neurological difference* (pp. 169–188). Minneapolis: University of Minnesota Press.
7. AASPIRE. (2013). About page. Retrieved from https://aaspire.org/about.
8. Nicolaidis, C., & Raymaker, D. M. (2015). Community based participatory research with communities defined by race, ethnicity, and disability: Translating theory to practice. In H. Bradbury (Ed.), *The SAGE handbook of action research* (pp. 167–179). Thousand Oaks, CA: Sage.
9. Nicolaidis, C., Raymaker, D. M., Katz, M., Oshwald, M., Goe, R., Leotti, S., et al. (2015). Participatory research to adapt measures of health and interpersonal violence for use by people with developmental disabilities. *Progress in Community Health Partnerships: Research, Education, and Action, 9*(2), 157–170.
10. Nicolaidis, C., Raymaker, D., McDonald, K., Dern, S., Ashkenazy, E., Boisclair, C., et al. (2011). Collaboration strategies in non-traditional CBPR partnerships: Lessons from an academic-community partnership with autistic self-advocates. *Progress in Community Health Partnerships: Research, Education, and Action, 5*(2), 143–150.
11. Nicolaidis, C., Raymaker, D., McDonald, K., Kapp, S., Weiner, M., Ashkenazy, E., et al. (2016). The development and evaluation of an online healthcare toolkit for autistic adults and their primary care providers. *Journal of General Internal Medicine, 31*(10), 1180–1189.
12. Stack, E. E., & McDonald, K. (2018). We are "both in charge, the academics and self-advocates": Empowerment in community-based participatory research. *Journal of Policy and Practice in Intellectual Disabilities, 15*(1), 80–89.
13. Nicolaidis, C. (2018, May 24). Hi, I'm Dr. Christina Nicolaidis and I'm editor of a brand new peer-reviewed journal called Autism in Adulthood. Ask me anything about the new journal or the ways that people on the autism spectrum can get better health care! (Online forum). Retrieved from https://www.reddit.com/r/askscience/comments/8lrqke/askscience_ama_series_hi_im_dr_christina.

14. Raymaker, D. M. (2017, February 27). I'm Dora Raymaker, an Assistant Research Professor at Portland State University, I conduct community-engaged research with the autistic and other disability communities. I am also autistic, and I am here today to talk about my research on autism and employment. AMA! [Online forum]. Retrieved from https://www.reddit.com/r/science/comments/5wj58r/science_ama_series_im_dora_raymaker_an_assistant.
15. Nicolaidis, C., Raymaker, D. M., Kapp, S. K., Baggs, A., Ashkenazi, E., McDonald, K. E., et al. (in press). Including autistic adults in research: Lessons learned and recommendations to the field. *Autism: The International Journal of Research and Practice.*
16. Hughes, R. B., Robinson-Whelen, S., Raymaker, D., Lund, E. M., Oschwald, M., Katz, M., et al. (in press). The relation of abuse to physical and psychological health in adults with developmental disabilities. *Disability and Health Journal.* Retrieved from https://www.sciencedirect.com/science/article/pii/S1936657418301948.
17. Oschwald, M., Nicolaidis, C., Raymaker, D. M., McCammon, M. A., Berlin, M., Joyce, A., et al. (2015). *Pregnancy, disability, and women's decisions* (Video series). Retrieved from https://pregnancyanddisability.org.
18. Oschwald, M., Renker, P., Hughes, R. B., Arthur, A., Powers, L. E., & Curry, M. A. (2009). Development of an accessible audio computer-assisted self-interview (A-CASI) to screen for abuse and provide safety strategies for women with disabilities. *Journal of Interpersonal Violence, 24*(5), 795–818.
19. Raymaker, D. M., Sale, T., Valera, M., Caruso, N., & Gould, V. (2018, March). *Empowerment of individuals experiencing early psychosis through community based participatory research and technology: Lessons learned from EASA connections.* Paper presented at the meeting of Annual Research & Policy Conference on Child, Adolescent, and Young Adult Behavioral Health, Tampa, FL.
20. Hunt, S. (2018, May). Fourth annual culturally responsive symposium highlights effective activism. *Vanguard.* Retrieved from http://psuvanguard.com.

Open Access This chapter is licensed under the terms of the Creative Commons Attribution 4.0 International License (http://creativecommons.org/licenses/by/4.0/), which permits use, sharing, adaptation, distribution and reproduction in any medium or format, as long as you give appropriate credit to the original author(s) and the source, provide a link to the Creative Commons license and indicate if changes were made.

The images or other third party material in this chapter are included in the chapter's Creative Commons license, unless indicated otherwise in a credit line to the material. If material is not included in the chapter's Creative Commons license and your intended use is not permitted by statutory regulation or exceeds the permitted use, you will need to obtain permission directly from the copyright holder.

11

Out of Searching Comes New Vibrance
Edited by Ren Stone

Sharon daVanport

My unwavering love affair with words led me down a melodic road more than once in my lifetime. Spending summer vacations reading dictionaries thrilled me as much as swinging from the vines in the woods near my childhood home.

And so it was in 2007 that I found myself staring curiously at my computer screen. I was captivated by a catchy turn of phrase that was foreign to me ("Nothing About Us Without Us")—and a word ("neurodiversity") that I had never read in any of the dictionaries I'd spent my formative years teething on. I continued gazing, as my synesthesia (where the stimulation of one sense can automatically lead to the stimulation of a second sense) and ideasthesia (where the activation of a concept can automatically lead to the perception of color) translated the new word and turn of phrase into several vibrant colors and harmonious sounds ("Types of Synesthesia" [1]).

S. daVanport (✉)
Washington, DC, USA
e-mail: sharon@awnnetwork.org

Earlier that same day, I attended an autism walk in my community. I was attempting to connect with other families and other autistic people. I didn't know much about the host organization other than the information I found on their website.

Later that evening, I went online to share how I had participated in my community's Autism Speaks walk. Within seconds, I received a private message from an acquaintance in the chatroom. With horror, I began reading a chilling account of an autistic girl named Jodie and her mother, Alison Singer, who was then a staff executive at Autism Speaks. Singer was featured in a documentary *Autism Every Day* [2], wherein she brazenly recounted that she had "contemplated putting Jodie in the car and driving off the George Washington Bridge"—all the while Jodie could be seen on camera playing in the background, within earshot of her mother.

Sickened by what I had just read, I immediately threw myself down a search engine rabbit hole. I found the video of Singer's interview and watched for myself *Autism Every Day* [2]. With utter disbelief, I viewed the film clip while struggling to swallow as I wiped the tears from my eyes. I couldn't wrap my mind around this mother's apparent lack of empathy and dismissal of her own disabled child. Singer, with a slight smile on her face, looks directly into the camera as she further admits that the only reason she didn't follow through with killing Jodie was because she had another daughter. I wondered how Singer found herself in an executive position of an autism organization if she was openly discussing thoughts of murdering her autistic daughter.

Nothing made sense.

Nothing.

I felt my fingers sweep across the keyboard as they took on a life of their own. My mouse clicked on link after link after link, and then something caught my eye. I found myself staring at a blue screen with small black font. I had stumbled upon the Autistic Self Advocacy Network (ASAN)'s website (autisticadvocacy.org). I began to relax as I absorbed the affirming, precise, and confident language that emphasized the importance of self-advocacy, activism, and a call to action. ASAN's message was the complete opposite of anything I had ever come across in the autism community to date. Instead of evoking feelings of doom and fear, I had felt a renewed sense of hope and empowerment.

Little did I know that my life was about to change. For the first time since entering online communities, I was reading messages about autism acceptance instead of the drivel of causation and cure. My mind was racing, and I ended up staying awake all night reading everything on ASAN's website as well as their linked articles.

I took several breaks throughout the night as I sat on my back porch piecing together recent conversations I had engaged in with my son which now became clear to me as the starting point on the path toward my understanding of autism acceptance as opposed to a cure. Everything transpiring that night seemed to be the culmination of my deepest desire to accept my son and myself as complete and whole people—perfectly imperfect, as all humans are in this world.

I had so much racing through my mind; how many more words expressed succinctly what I would experience in this flood of feelings? The euphoria was drowning me…and then I was sitting; that catchy turn of phrase and new locution was right there in front of me. Everything I had read that night on the ASAN's website ran circles through my mind while I read kept returning to this newly discovered quote, "Nothing About Us Without Us" and this new word: neurodiversity.

It all rang true; it felt right. Learning about neurodiversity and the importance of autistic people being part of all conversations and every decision involving their lives made absolute perfect sense. So much so that it was hard to grasp that I had never thought of their importance before that night.

Over the next few months, I often struggled to digest the onslaught of information I was taking in. Sometimes the activist bloggers I followed felt loud and I sensed when their energy would trigger my PTSD. All the same, I appreciated the importance of their work and continued to push forward. I knew that no matter how tough it was to hear the experiences of autistic people fighting for their rights, I needed to understand. I would at times get overwhelmed and go offline for up to a week to process what I was learning. Then I'd return to take in as much as I could before I needed another break.

I learned quickly that I had spent my life in ignorance as it pertains to the injustices experienced by people with disabilities. The more I researched, the more I discovered the importance of social justice activism and its vital place in the world.

Out of Searching Came Community

Neurodiversity soon became something that I intimately understood as the all-inclusive acceptance of every neurological difference without exception.

I further came to appreciate that neurodiversity didn't leave anyone out. Even the opponents of this concept reaped the benefits of the work by neurodiversity activists. It didn't matter whether they agreed with the concept or not, they personally benefited. Furthermore, their children did as well, as the specific premise of neurodiversity is full and equal inclusion.

It wasn't too long before I reached out to the ASAN, and over the next couple of years I developed friendships with ASAN's founders, Ari Ne'eman and Scott Robertson.

One of the most influential youth activists for disability rights during that time was Savannah Logsdon Breakstone. Savannah and I developed a close friendship which eventually led to the ideas that ultimately influenced the beginning of Autistic Women and Nonbinary Network (AWN) in January 2009.

One of the greatest life lessons that I learned during the early years of my involvement in the autistic community is that we are no different than any other community of people; though we are linked by a familiar neurology, we are still individuals in our own right, with differing opinions, contrasting ideas, and conflicting access needs. When you put all of that together with different personalities, it makes for a brilliantly vibrant and sometimes challenging community.

By 2008, I had developed friendships with several other autistic women through online groups. Savannah and I, as well as many of our friends found it difficult to fit into many of these groups, and we found ourselves searching for a community of women with powerful, balanced, and non-hierarchical leadership that shared our core beliefs of autism acceptance and disability rights, as well as an understanding for increased advocacy and resources as it relates to autistic women and the gender disparities they face.

After several more emails of encouragement from Savannah, I decided to take the plunge. I contacted one of my good friends and web developer, Lori Berkowitz of BeeDragon Web Services (beedragon.com) to discuss the possibility of forming an online community and forum for autistic women. Lori and I had gotten to know one another over the past year through our involvement in another group, and we had also worked together briefly on another website project.

Within a few short months, Lori had the website and forum ready to launch. Autistic Women & Nonbinary Network officially went online in January of 2009 (awnnetwork.org). Those early days were full of nervous energy. Ari Ne'eman of ASAN was a huge help with offering tips and guidance—do's and don'ts—when creating an online presence.

I reached out to several spectacular autistic women to form Autistic Women & Nonbinary Network's founding board of directors: Sandy Yim, Tricia Kenney, Lindsey Nebeker, Savannah Logsdon Breakstone, Corina Becker, and Lori Berkowitz were AWN's founding superstars in the early years. What we originally perceived as an online forum type of community quickly grew larger than we ever imagined.

In 2009, Tricia Kenney and I created and hosted AWN's BlogTalk Radio Show (blogtalkradio.com/autism-womens-network), and by 2010, Autistic Women & Nonbinary Network had gained national recognition. In April, I was invited to represent AWN at a White House meeting on World Autism Day, and to participate in discussions with President Barack Obama's Administration related to their ongoing efforts to better support autistic individuals. In July, I was invited back to Washington, DC for the White House's 20th Anniversary Celebration of the Americans with Disabilities Act.

There seemed to be no letup with how quickly Autistic Women & Nonbinary Network was growing. Autism research has historically been focused on young school-aged boys, and it was becoming increasingly evident that AWN was fulfilling a great need in our community by keeping the focus on advocacy, resources, and research for autistic women and girls and non-binary people.

AWN steadily became a regular participant in national autistic-centered conversations, and we began seeking opportunities to widen our intersectional activism. I am of the firm belief that a decision-making board or supervisory board's composition should be representative in gender and race. Once a predominantly white cis board of directors, we as a board recognized our need to change and have been intentional about making sure representation is diverse because that is what makes a strong organization. AWN's current board representation is inclusive with intersectional diverse leadership. For the last three years (as of 2019), the board has consisted of mostly people of color and has always been majority LGBTQ+ members. To speak more directly to that intersectional diversity, in 2018, we changed our name from Autism Women's Network to Autistic Women & Nonbinary Network.

From its inception, Autistic Women & Nonbinary Network has strived to provide our board members with personal accommodation requests. This ultimately led to all board meetings being conducted solely online, and since 2012 our meetings have been exclusively held via text-based/real-time communications.

AWN sees itself as part of the wider disability rights movement. Undeterred in our quest to fight for autistic rights, we appreciate our place as being part of the greater civil rights movement for disabled people. In this spirit, we are currently working on leadership development for our committee for our initiative Divergent: when disability and feminism collide (facebook.com/DivergentFeminism).

> Divergent works to change how disabled women are commonly perceived within society while challenging the myths of our inferiority, both as women and as disabled people. We explore the interactions between sexism and ableism within both disabled and nondisabled communities. We seek to offer perspectives on gender and disability by emphasizing non-traditional femininity and non-traditional feminism. [3]

Reflecting on the years gone by, I can't help but feel immense gratitude for all the people and experiences along the way. Time has a magical way of bringing about clarity of purpose. The impact of this clarified praxis opened conversations around the greater disability community, helping us as an organization to expand our mission. Most recently, we learned in March 2019 that Lori's tireless work to build our online platform had led to recognition by the United States Library of Congress as a culturally significant contribution to society and our content is now archived nationally. The Library of Congress states that AWN's website has been selected "for inclusion in the Library's historic collection of Internet materials related to the Women's and Gender Studies Web Archive" as they consider our website to be an "important part of this collection and the historical record."

Even with all the progress made over the years with respect to disability rights (and specific to autism) we still live in a lopsided world which measures a person's worth based upon false premises. Ableist rhetoric taints the conversations which lead to discrimination. And still, here we are, Advocates and Activists, with even more vigor and determination in the face of all that has dared to silence us. Despite the uneven roads we often travel, we know without a doubt that we will not be erased. It's been an amazing journey so far, and I'm looking forward to the years ahead and experiences yet to come (Fig. 11.1).

NEURODIVERSITY IS FOR EVERYONE
autistic women & nonbinary network

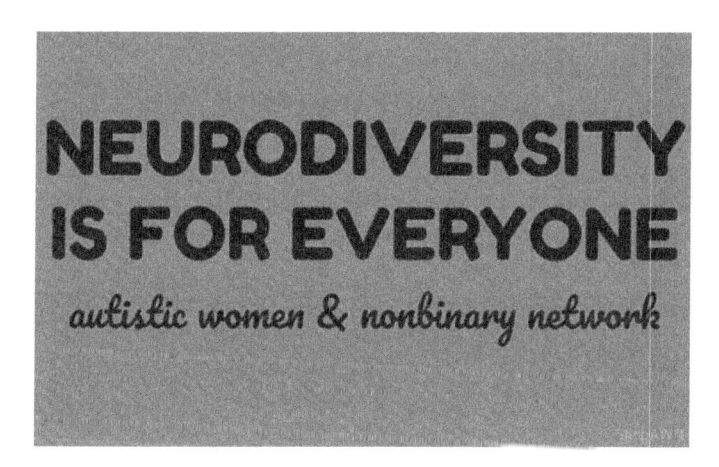

Fig. 11.1 Autistic Women and Non-Binary Network tagline

Neurodiversity is for everyone

Nothing About Us Without Us!

References

1. Synesthesia.com. (n.d.). *Types of synesthesia*. Retrieved from https://synesthesia.com/blog/types-of-synesthesia.
2. Watkins, J., Solomon, E., & Thierry, L. (2006). *Autism every day* (Documentary). USA: Milestone Video, The October Group.
3. Liebowitz, C. (2013, December 17). *Divergent: When disability and feminism collide* (Web log post). Retrieved from https://awnnetwork.org/divergent-when-disability-and-feminism-collide.

Open Access This chapter is licensed under the terms of the Creative Commons Attribution 4.0 International License (http://creativecommons.org/licenses/by/4.0/), which permits use, sharing, adaptation, distribution and reproduction in any medium or format, as long as you give appropriate credit to the original author(s) and the source, provide a link to the Creative Commons license and indicate if changes were made.

The images or other third party material in this chapter are included in the chapter's Creative Commons license, unless indicated otherwise in a credit line to the material. If material is not included in the chapter's Creative Commons license and your intended use is not permitted by statutory regulation or exceeds the permitted use, you will need to obtain permission directly from the copyright holder.

12

Two Winding Parent Paths to Neurodiversity Advocacy

Carol Greenburg and Shannon Des Roches Rosa

Shannon Des Roches Rosa: Encountering Neurodiversity as a Terrified Outsider

I came into neurodiversity activism sideways. It isn't the main reason I co-founded the online nexus and book *Thinking Person's Guide to Autism* [1]—the goal at the time was to debunk autism misinformation, and provide useful, evidence-based resources for autistic people, their families, and autism professionals—but neurodiversity, with its respect for and celebration of diverse neurologies, ended up being our organization's guiding principle.

I didn't know the word "neurodiversity" in 2003, when my son was diagnosed with autism. I wish I had; maybe I wouldn't have fallen so hard for misinformation about vaccine, causation, and autism cures. I admit

---

C. Greenburg (✉)
New York City, NY, USA

S. D. R. Rosa
Redwood City, CA, USA

© The Author(s) 2020
S. K. Kapp (ed.), *Autistic Community and the Neurodiversity Movement*,
https://doi.org/10.1007/978-981-13-8437-0_12

that being exposed to that era's sensationalistic and negative media messages about autism made me vulnerable to false-hope-based cure hawkers, and that I put my son through "treatments" that were a waste of time and money.

I will never stop being ashamed of how, under the guidance of a medical doctor who convinced me and my husband that he could "treat" autism, I subjected my autistic preschooler to a full autism quackery barrage: innumerable supplements and dietary restrictions, pseudoscience "electrical field" treatments, vitamin B12 injections, and even preparation for chelation—all of which I publicly detailed at my personal blog squidalicious.com ([2]; though I was writing under the pseudonym "Squid Rosenberg" at the time).

Rejecting Autism Quackery

Though it took too long, I eventually wised up to the fact that no child deserves to be treated as a fixer-upper rather than a fully present human being—especially the sweet little boy I was supposed to be fighting for. I will always be grateful for the frank talking-to my medical professional father-in-law gave my husband and me on not subjecting our son to chelation "treatment", as well as my guidance from generous science- and neurodiversity-minded individuals about autism origins and autistic ways of being. The only way to pay back that debt is to pay it forward.

Once I realized that we had been not only duped but fleeced by autism quacks, and were making my son miserable while autistic rather than non-autistic, since he was born with his autistic brain, I became hellbent on helping others avoid my very avoidable mistakes. I shared my autism and parenting epiphanies publicly, not only on my blog, but in Steve Silberman's book *NeuroTribes* [3]—mortifying though it remains to detail my early fear and ignorance regarding my own child's needs.

I also founded Thinking Person's Guide to Autism (TPGA) in 2010 with Jennifer Byde Myers, Emily Willingham, and Liz Ditz. We set about gathering evidence-based autism resources presented in straightforward but supportive terms, addressing some of our community's starker realities, as well as barriers to accessing services and accommodations. We started by publishing articles on our website, with the best going into our 2011 book. We also built vibrant social media communities.

Learning Why Neurodiversity Matters from Insider Perspectives

One of our first published essays, sourced by Ditz, was Mike Stanton's "What Is Neurodiversity?" [4]—also one of my first exposures to the term. Stanton emphasized the need to respect autistic ways of learning and perception as not just "different" but legitimate. We also began working with more autistic authors, and realized having an autism organization with no diagnosed autistic team members was both inappropriate and embarrassing. We invited Autism Women's Network officer Carol Greenburg, a skilled editor as well as an autistic mother of an autistic son, to join our team. Thankfully she agreed to do so, to TPGA's ongoing benefit as well as mine—Carol has become one of my dearest friends, in addition to being a treasured colleague.

As TPGA grew and expanded, and our author base skewed increasingly autistic, my neurodiversity education grew, along with that of many of our community members: as an autistic community outsider I can't understand the autistic experience without access to autistic insights, and neither can other non-autistic community members—especially since autistic experiences, traits, perspectives, abilities, and personalities are multitudinous, even as autistic commonalities unite the community.

I have become reliant on neurodiversity-informed perspectives, on the insights of people who describe the reality that my mostly non-speaking autistic teenage son and other autistic people experience, and actually help make the future they deserve to happen. Without the neurodiversity concepts of respecting and supporting different minds and abilities, without the inclusiveness neurodiversity demands, autism advocacy efforts risk becoming factionalized and leaving people in need without the support and community they deserve.

Though essays like Julia Bascom's "Dear 'Autism Parents'" [5] initially felt hostile, with statements like "If you do indeed, as you claim, want to be *allies*, then I suggest you start *acting like it*," when viewed through a neurodiversity lens, it is simply a statement of affirming autistic identity and rights—and for my son, too. When I watched Mel Baggs's video, "In My Language" [6], and saw that Mel is multiply disabled in ways mirroring my son yet could also describe that experience in writing, it destroyed my unexamined assumptions about the boundaries between disability and ability. And when the Autistic Self Advocacy Network (www.autisticadvocacy.org) demands "nothing about us without us," they are taking their rightful seat at the head of the autism roundtable, rather than trying to bar the concerns and questions of non-autistic parents like me from those discussions entirely.

I also had to confront flaws in my own thinking as my understanding of neurodiversity evolved. I fought against autism-vaccine disinformation because such statements are provably false. But neurodiversity activists like Emily Paige Ballou (chavisory.wordpress.com) helped me understand that not only do vaccines "have nothing to do with autism," but framing autism as an "injury," or even saying "don't worry, vaccines don't cause autism" is still making my son's neurology into something to be feared rather than understood and accepted—and that is both stigmatizing and counterproductive.

Some neurodiversity-informed approaches have been harder to sow, due to widespread misinformation about autism and disability. We're still working on helping people understand that it doesn't matter how many people (and industries and lobbyists) support and promote applied behavioral analysis as a "treatment" for autism, if the goal is to turn an autistic child into a non-autistic one, because that's an approach many adults who experienced ABA as children describe as PTSD-inducing. Instead, we need more people to understand and work with autistic children's unique set of intellectual, visual, sensory, auditory, communication, and motor processing abilities.

Overall, I believe the team efforts behind TPGA have been successful. We hear from people every day who are grateful that we include perspectives that reflect their own experiences, demonstrate why presuming competence matters, further the shaking off of stigma, make them feel less alone, and provide insights into their children, students, clients, or even selves that they'd never considered before. And most of these efforts wouldn't be as useful as they are, if they weren't neurodiversity-informed.

I will always be learning more about what neurodiversity means. I now better understand the foundational drive of respecting and ensuring rights for people of diverse neurologies, and reject my initial assumption that neurodiversity advocates deny disability, but the process will always be a journey. (And not always an easy one: that while neurodiversity advocates are often the most delightful and compassionate people one will ever meet, neurodivergence doesn't protect autistic people from being capable of flawed logic, bad choices, ignorance, and outright cruelty.) But I now understand that neurodiversity is not only a fact of life but a litmus test: I can't trust people who don't support neurodiversity, not when it comes to autism best practices, or my son's future.

Carol Greenburg: Advocacy Rooted in a Neurodivergent Family Tree

Before I knew many autistic adults who knew they were autistic, before I met any autistic bloggers, before TPGA existed and asked me to write a piece for them, it was about the money for me. I had helped a friend to get services for her child on the spectrum and realized I had a knack for cutting through the layers of red-tape that cordoned children like my son off from services to which they were legally entitled. I thought I could do some good and get paid for it.

So I decided to become a non-attorney special education advocate, and get paid for my work. Non-attorney advocates don't, and shouldn't charge anywhere near as much as people who go through law school and pass the bar, but even without a legal degree helping families properly prepare for and effectively represent them at IEP meetings takes knowledge and attention to detail. Processing that knowledge takes time and barrels of energy, which I think should be reflected in a paycheck.

I see the option of paid work, of contributing to my household income, as a part of adulthood I had never quite mastered. I was tired of the message that getting paid consistently with real cash for my real expertise was not a goal I was likely to reach. My first clue? Chronic under employment and unemployment in my previous attempts at paid labor, the annoying sticky companion of so many autistics, so many women, and especially autistic women like me.

Suddenly My World Made Sense

Despite my education and skills, I was fired from almost every job I held in my twenties and into the first half of my thirties. I didn't understand why, until what I had suspected was confirmed by the autism spectrum diagnosis I got when I was 44. The notion that I am autistic didn't surprise me at all. Autism doesn't just run in my family, it gallops. My son was diagnosed at 3 1/2. I called my Dad the day after we learned about my son's autism; back then Dad could still use a phone. Absurd as it now seems to me, I struggled a bit, probably more with my own internalized ableism than with any rational fear of his reaction to the question I wanted to ask him: Did he think, as I did, that there was a reason for the peculiar similarities between the three of us? To his eternal credit he didn't laugh at me, but to this day I maintain I could practically hear him roll his eyes over the phone when he gave me a definitive yes.

My Dad never got a formal diagnosis, but even before he lost most of his spoken language to a series of strokes, he has quietly carried an autistic pride banner for all of the rest of our autistic family, living and dead. In communications between the two of us, ever more halting and difficult as they are to conduct, he has expressed our shared belief that we would not trade our autistic reality for some artificial construct invented by non-autistics whose brains and experiences differ from ours so much. I did and do still check in with him periodically to make sure it's OK that I've been continually outing him as autistic and proud for over a decade. He used to tell me that he felt if I could spread that message I should, for the sake of all of us, as he dislikes how we are portrayed as having lives not worth living. Nowadays, he gives me an elaborate flourish of his hand that says to me "Carry on!"

My diagnosis took the edge off that shame and gave me some valuable context: It was now clear how much my autistic brain shaped my autistic cultural assumptions. Office politics were not only senseless, but actively offensive, built as they were on a scaffold of lies and tacky dominance performance art. I wouldn't have participated in them even if I could have. That was only one of many social trappings of standard workplaces from which I hoped entrepreneurship would free me. The problem became not so much what, as how: I knew I wanted to start my own company as a non-attorney special education advocate. The problem was, how would I attract clients?

I had written professionally before the Internet existed, so I had some sense of what writing required, and now with the existence of blogs, I had a ready-made platform. My blogging didn't pay, but I thought my writing might help establish my expertise for clients I hoped to attract. I saw the most positive responses I got as contributing to good Karma if not immediately to my piggy bank, and tried to keep faith in the notion that getting my name out there would eventually yield some concrete results.

Paid or unpaid, advocacy by autistics for autistics is vital, but when I started this work, I felt strongly that my brand of advocacy would be linked to the economics of womanhood and autism. I was tired of lectures about the beauty of voluntarism. Women's work, autistics' work seemed to result in strikingly similar outcomes: enthusiastic praise that was supposed to somehow compensate for the lack of financial parity. Service-for-service's-sake is a principal that can stretch only so far before it becomes a burden bulging with the resentment of those serving.

However the eternal problem with a do-over in a new field still hovered over my goal. It is difficult to begin a new work life without taking a few steps backward, which in my case meant more volunteering in addition to writing. I ended up creating a DIY apprenticeship to settle myself into presenting myself as a professional. I started helping longtime friends with their children's IEP prep and meetings. In between helping individual families, I wrote, spoke, designed and delivered workshops, almost always for free. Sometimes, I had great experiences: my first clients, old friends all, treated me with respect because that's how friends act.

The treatment I got from people I didn't know, the ones who wanted me to do free workshops or write free articles: that was more uneven. Although some of these strangers eventually became welcome, enduring parts of my world, others I couldn't escape fast enough.

One workshop organizer scolded me for using the word "Autistic" to describe myself—and my own son. To her, I had no right to describe my own personhood in the terms of my choice. Workshop participants, mostly non-autistic parents of autistic kids, often meant well, but did no better than that organizer, and thought they were complimenting me when they said I didn't "seem autistic" like their and my own mostly non-speaking kids (I speak pretty fluently, unless some upset robs me of language). I felt that thanking them for their attempts to praise me violated my personal policy of not accepting compliments at the expense of my people: It never makes me happy to hear that I'm not like other autistics—or women, or feminists, or Jews—anyone has met.

Then there were the editors, who asked for free articles, which they wanted to rewrite to reflect their own usually non-autistic biases. I like my own always-autistic biases better though, and except for minor edits for clarity, I held firm to my belief in my ability to spout my own opinions better than anyone else could spout them. I was getting discouraged. Then a new blog called Thinking Person's Autism asked me to submit a piece and my world changed again.

A Chance Encounter. A Permanent Change

A month or two after I wrote that piece for TPGA, a chance encounter became a permanent part of my life. It wasn't a paying gig, but it has led to many, and at least as importantly: I finally got to get used to the kind of treatment I think all people deserve. At a panel at a BlogHer conference about busting autism myths, I met the woman who came up with the topic for the panel, Shannon Des Roches Rosa. She was there with Jennifer Byde Myers, her co-founder of the new project TPGA. Shannon had an idea that should seem pretty obvious to all of us now, but Simply Wasn't Done back then: She invited an Actual Autistic to join the panel.

As it turned out, the wonderful autistic advocate they originally invited was unable to attend. I volunteered to pinch-hit last-minute. I met Shannon, Jen, and the other panelists ten minutes before we went on.

I'll never forget the first thing Shannon said to me: She asked if I needed help getting up to the podium. I was amazed by the thoughtfulness of the offer. Like many autistics, I have trouble with proprioception, the awareness of where my body is in space. Although I can use stairs without assistance, and if I try hard enough to mask the effort, I can hide how hard it is for me to just get around without falling, bumping into objects and people, or knocking things over. But when I have a problem that indicates my impairments, I'm used to people laughing at me, despite my explanations—because I just don't seem THAT autistic. So the only logical conclusion is that I'm clumsy, or I'm not paying attention or—as any autistic reading this has undoubtedly already predicted—I'm not Trying Hard Enough. Even though I am trying as hard as I can, every moment of every day of my life, even during the moments only I can see and feel. Shannon got that instinctively.

Later that day, Jen offered to let me wait in her hotel room while she grabbed something she needed for an upcoming presentation. I hesitated, because I have experienced that wrath of those who have made insincere, token offers that I accepted because I didn't know some rule (hidden to me) that I was supposed to demure. I decided to take a chance and explain that she needed to be absolutely clear with me about whether her offer was genuine, because I need spoken subtitles for social niceties. I realized her offer was real when she actually thanked me! No stranger for whom I had done a workshop or panel had ever thanked me for telling them exactly what kind of accommodation I needed. Mostly they just seemed annoyed.

TPGA: A Neurodiverse Ecosystem

A few months after meeting Shannon and Jen, I was honored by an invitation to join the editorial team of TPGA. We pay our authors, but I don't get paid for my TPGA editorial work; none of the editors do. My opinions carry the same weight as those of the non-autistic editors. I want to emphasize that: The Same Weight. So if our audience likes what we publish, I may play some part in our having published it. Conversely, if they loathe what we publish, I bear as much responsibility for it as anyone else on the editorial team.

My work with TPGA remains one of the great honors of my life. Staff discussions and occasional disagreements may be blunt, but they are also respectful. While I can't define the totality of Neurodiversity even for myself, much less anyone else, I know spaces where I can see it. TPGA is had always been one of those spaces for me.

I still do a lot of public speaking, often for money and/or at least travel expenses, and I enjoy it. How I am introduced depends upon the organization and audience. Some autism professionals and parents still feel uncomfortable around adult autistics, perhaps they expect me to judge their efforts harshly, so they soft-pedal my own autism and play up my role as the parent of an autistic son.

When I speak publicly with my co-editors at TPGA, I think it is not at all coincidental when whoever has invited us makes the greatest effort to combine all of my roles in the autism community. Those are the times when the word "self-advocate" pops up, and I do understand why, but it makes me a little uncomfortable. It is a term many autistics embrace, but I don't find it entirely descriptive of what I do, professionally or personally. Do I self-advocate and teach self-advocacy to my autistic son? Of course, because self-advocacy means nothing more or less than insisting on one's full rights. In my view self-advocacy is not a job in and of itself, but an expression of our dedication to living as we are. For people with disabilities in particular, the world does not approve of us accepting ourselves as is.

With the death of some in my family, and other circumstances in the lives of the other autistics whose light we can still see in this world, I'm the only autistic left who uses spoken and written languages with any fluency. Of course, I'm active in protecting my father's and son's rights, and I am accustomed to but continually disgusted by the efforts of many non-autistics to treat them as somehow lesser because they don't express themselves primarily with written or spoken language.

Like all loved ones my father and my son are sometimes sources of frustration and sometimes sources of joy, but that's a function of their humanity, not their disability. I have never been and will never be the only person in my family who has expressed agency in my own life, and my allegiance to my autistic family members' rights stands, no matter what. Just I see myself as a full human being with all of my human rights intact I ally with my autistic family members who have fewer words when they get subpar treatment, because I am a daughter and mother of disabled men who are equally, fully human.

That is where Neurodiversity starts to gel—at least for me. Neurodiversity in my world, is the unquestioned right for all, whatever their neurological makeup, to express what they need or want. None of us get what we need or want all of the time, and that's OK. Getting everything you request, or demand, is not the reality of anyone I know, no matter how much privilege they have, whether they own up to it or not. I don't think any of us, of any neurostatus, should get every golden, silver, or copper ring we reach for, but everyone should have stable, level ground from which we can launch authentic discussions about what we owe and what we deserve.

Conclusion

We wouldn't know each other if it wasn't for neurodiversity, and our individual journeys in understanding and embracing what that means both for ourselves and for our families. Now that our paths have crossed, we are obligated to collaborate on sharing our ever-increasing knowledge with the world. We also have the (often humbling) experience of hearing how our work, especially with TPGA, is positively influencing others' lives—autistic people and their families, of course, yet also professionals, academics, policy makers, and researchers. I hope we'll continue to working together for the foreseeable future, because opening people's minds to why neurodiversity matters—both as a human rights concept and because individuals deserve respect—is a worthy endeavor.

References

1. Rosa, S. D., Myers, J. B., Ditz, L., Willingham, E., & Greenburg, C. (2011). *Thinking person's guide to autism.* Redwood City, CA: Deadwood City Publishing.
2. Rosa, S. (2003, September 12). *Look! Ten apples up on top! I am so good I will not stop!* (Web log post). Retrieved from http://www.squidalicious.com/2003/09/look-ten-apples-up-on-top-i-am-so-good.html.

3. Silberman, S. (2015). *Neurotribes: The legacy of autism and the future of neurodiversity*. New York: Penguin.
4. Stanton, M. (2010, June 17). *What is neurodiversity?* (Web log post). Retrieved from http://www.thinkingautismguide.com/2010/06/what-is-neurodiversity.html.
5. Bascom, J. (2011, August 23). *Dear "autism parents"* (Web log post). Retrieved from https://juststimming.wordpress.com/2011/08/23/dear-autism-parents.
6. Baggs, M. (2007, January 14). *In my language* (Video file). Retrieved from https://www.youtube.com/watch?v=JnylM1hI2jc.

Open Access This chapter is licensed under the terms of the Creative Commons Attribution 4.0 International License (http://creativecommons.org/licenses/by/4.0/), which permits use, sharing, adaptation, distribution and reproduction in any medium or format, as long as you give appropriate credit to the original author(s) and the source, provide a link to the Creative Commons license and indicate if changes were made.

The images or other third party material in this chapter are included in the chapter's Creative Commons license, unless indicated otherwise in a credit line to the material. If material is not included in the chapter's Creative Commons license and your intended use is not permitted by statutory regulation or exceeds the permitted use, you will need to obtain permission directly from the copyright holder.

13

Lobbying Autism's Diagnostic Revision in the DSM-5

Steven K. Kapp and Ari Ne'eman

Defining the boundaries of autism has always been a complex task, shaped by a wide variety of scientific, social, political, and economic factors. Those boundaries shape the lives of autistic people, influencing not only who gets diagnosed but often providing significant and important context to clinical decisions about service provision and "treatment" along with setting the stage for lifelong diagnostic and service disparities on the basis of gender, race, class, and age.

S. K. Kapp (✉)
College of Social Sciences and International Studies, University of Exeter, Exeter, UK
e-mail: steven.kapp@port.ac.uk

Department of Psychology, University of Portsmouth, Portsmouth, UK

A. Ne'eman
Senior Research Associate, Harvard Law School Project on Disability, Cambridge, MA, USA

Because autistic people are shaped by the diagnostic process, one of the Autistic Self Advocacy Network's priorities—as the leading organization run by and for autistic people—was to shape that process in return. We sought to do this with a variety of goals in mind: to address existing diagnostic disparities, improve access to service provision where diagnostic distinctions interfered, and to prevent a loss in access to legal protections, social legitimacy, and service provision by the narrowing of the diagnosis. While the Neurodevelopmental Disorders Workgroup charged with revising the autism diagnosis in the Diagnostic and Statistical Manual of Mental Disorders, Fifth Edition (DSM-5; American Psychiatric Association [1]) consisted of researchers who conducted analyses and whose decisions received reviews from academic scholars [2], the process was still a political one, subject to efforts to influence the outcome. As a representative acknowledged, "This is not science – this is a committee" [3]. Furthermore, we maintain that the scientific and research processes are framed and mediated by larger social and political ones, and thus that dedicated advocacy and lobbying could influence the resulting diagnosis. In this, we were absolutely correct.

ASAN's advocacy work regarding the DSM-5 was led by the two authors, Ari Ne'eman (ASAN's co-founder and then President) and Dr. Steven K. Kapp (then a doctoral student at the University of California, Los Angeles and ASAN chapter Co-Director). While the organization was pursuing political and policy goals, we sought to ensure that ASAN's advocacy would be well-grounded in the research literature so as to maximize the likelihood of success and ensure the organization's credibility.

Ari led the lobbying effort and served as the primary point of contact with members of the DSM-5 Neurodevelopmental Disorders Workgroup. He also served as the primary expert on law and policy considerations in service provision. Steven led the research expertise side, serving as ASAN's technical expert on the research literature, providing comprehensive information on the existing autism research literature, and ensuring that the organization was capable of responding rapidly to questions or concerns raised by Workgroup members regarding the research literature.

Larger Context of Diagnostic Process

While the DSM had been revised previously, the current diagnostic process took on outsized public attention for a variety of reasons. Some of this was due to the simple fact that during the development of the DSM-IV [4], an organized community of autistic adults did not yet exist in significant numbers. The DSM-5 was the newly organized autistic community's first opportunity to weigh in on the criteria that governed who the medical community considered autistic.

But the DSM-5 process attracted additional attention for another reason: many in the autistic and autism communities were gravely concerned by rhetoric that autism was "over-diagnosed". Though the expansion of the diagnostic criteria in the DSM-IV had given large numbers of people access to legal protections, service provision, and a diagnosis and communal identity that helped them make sense of lifelong experiences of social isolation, odd interests, and other common autistic experiences, it had also sparked a backlash among some clinicians and members of the general public. Early media reports about the DSM-5 process suggested potential intent to narrow the diagnostic criteria [5]. These reports noted that the pathways to an autism spectrum diagnosis would shrink from 2027 to 11 possible "symptom" combinations [6] and that the committee had laid out an official goal to avoid false positives [5]. Further reports that the proposal would narrow the criteria significantly [7] sparked anxiety and deep worry among many. While the Workgroup did have another goal of improving identification in women and girls, racial and ethnic minorities, and adults—admitting the DSM-IV worked best for five-to-eight-year-old white boys [8]—further reports that the proposal would narrow the criteria significantly [7] sparked anxiety and deep worry among many that the proposal would leave many without access to the diagnosis who might benefit from it.

The committee's early proposal to combine the DSM-IV's main three autism diagnosis, Autistic Disorder, Asperger's Syndrome, and PDD-NOS, into a single unified autism diagnosis exacerbated these fears (though it was not the origin of them, e.g. Giles [9]). Many autistic people opposed the integration of the Asperger's diagnosis in particular into the larger autism spectrum.

However, this proposal was not intended as a measure to narrow the scope of the autism spectrum. Instead, it was rooted in significant research and clinical findings that the three autism diagnoses were applied inconsistently depending on the age and background of the person being diagnosed and the physician conducting the diagnosis [10]. Many individuals would receive multiple autism diagnoses across their lifespan, reflecting the fact that the three diagnoses had come to be used as a proxy for quality of outcome rather than being reflective of different phenotypes of autism. To quote one early commentary by the Neurodevelopmental Disorders Workgroup, "A single spectrum disorder [i.e., folding in Asperger Disorder and PDD-NOS] is a better reflection of the state of knowledge about pathology and clinical presentation; previously, the criteria were equivalent to trying to "cleave meatloaf at the joints" [11].

The proposal to unite the autism diagnoses paradoxically divided the autistic community, with individuals diagnosed with, and organizations based on, Asperger's leading the opposition. Michael John Carley, executive director of the Global and Regional Asperger Syndrome Partnership (GRASP, then led by individuals with the Asperger's diagnosis), represented the sense of superiority many of these critics felt over autistic people with higher support needs. "I personally am probably going to have a very hard time calling myself autistic," said Carley in an interview with National Public Radio, comparing the cultural perception of Asperger's as a diagnosis perceived as associated with major historical figures, like Edison and Einstein, to "somebody who might have to wear adult diapers and maybe a head-restraining device. This is very hard for us to swallow," [12]. While Carley [13] couched GRASP's leadership of the opposition to the DSM-5 in terms of concerns about diagnostic narrowing, he thus initially voiced his personal discomfort with the removal of a separate Asperger's diagnosis based on cultural identity. ASAN did not share this worldview. While we recognized that "autism" carried with it more stigmatized connotations than "Asperger's", we believed that such stigma could be changed. More importantly, there was no valid reason why it should be concentrated toward only one part of the autism spectrum until such time as that change could be accomplished. Though both Ari and Steven possessed Asperger's diagnoses, it was our belief that the best way to address stigma was to confront it across the spectrum. Why did we deserve protection that other autistic people did not receive?

This was both reflective of our commitment to "cross-spectrum solidarity" and the essentially arbitrary process by which one individual might receive a particular diagnosis while another similar individual might receive another. Though "Aspie Supremacism" had been a longstanding problem in certain circles of the autistic community [14], ASAN had always insisted on a cross-spectrum perspective and consisted of a leadership and membership drawn from individuals who had received all three of the diagnoses (as well as some who had been unable to access a diagnosis due to various disparities).

We also believed that the three separate diagnoses contributed to service eligibility gaps, where laws, regulations, and policies by payers provided for eligibility for those with one diagnosis but not for others with comparable levels of impairment and need. In addition to their lack of clinical and research validity, ASAN had documented numerous instances where the three different diagnoses were used to limit access to services.

But concerns remained that, if the DSM-5 was implemented in an insufficiently precise fashion, some would be pushed out of the diagnosis. Early research on DSM-5 draft proposals suggested that the revision might lead to a narrowing in the availability of a diagnosis, pointing to early estimates that predicted a severe consolidation of as much as 54% overall (100% for those with Asperger's diagnoses in their sample; [15]). Asperger's had been crucial to the broadening of the eligibility for an autism diagnosis when the DSM-IV had come out, and many who had gained access to diagnostic legitimacy, legal protections and service provision feared their loss [9]. While ASAN supported the shift to a single unified diagnosis, we shared those concerns and engaged in advocacy in part to protect members of our community against the harms associated with the loss of a diagnosis by advocating for a broad formulation of a unified diagnostic criteria.

ASAN also sought to use the DSM-5 process to address other equity concerns, specifically race, gender, age, class, and geographic disparities in access to diagnosis. Significant racial disparities in access to diagnosis and service provision had been documented, with African-American and Hispanic children less likely to receive a diagnosis and, among those that did, the diagnosis typically came later in life and for those individuals with more "severe"—obvious—autistic traits [16].

Similar gaps existed with respect to gender, though these disparities were often constructed as real biological facts rather than disparities in access to diagnosis. However, the autistic community had long maintained that, while the actual rate of autistic men and boys to autistic women and girls could not be definitively known, a significant percentage of that gap was attributable to gender bias and the resulting disparities. A growing body of research literature was coming to agree with us [16]. Furthermore, ASAN maintained that the DSM-IV criteria often made it difficult for autistic adults to receive a diagnosis, since we tended to develop various "masking" or "passing" skills as we grew up that hid the autistic traits we had had in childhood, even as the effort associated with passing still created cognitive demands and quality of life challenges not experienced by non-autistic persons [16].

Finally, we were deeply worried about proposals to write into the DSM-5 criteria for "recovery", reflective of a small number of studies that claimed to show autistic children losing their diagnosis in adulthood or adolescence. ASAN was skeptical of these findings, as a number of our members had been deemed "recovered" in childhood only to be re-diagnosed or find the autism diagnosis of continued relevance to them in adulthood. Even within the research literature supporting recovery, the vast majority who "lose" an autism diagnosis had it replaced with another diagnosis and continued to face significant challenges associated with the autism spectrum, suggesting that they were in fact simply learning how to "pass" and develop coping skills [17, 18]. ASAN was concerned that writing "recovery" parameters into the DSM-5 autism criteria would result in individuals losing their diagnosis and resulting access to services, legal protections, and communal identity when they develop meaningful coping mechanisms.

As a result, we advocated for the DSM-5 workgroup to avoid "recovery" criteria and to write into the DSM-5 autism diagnosis that individuals could be diagnosed based on present or past manifestations of autistic traits. Specifically, we sought to codify that learned behavior or other "mitigating measures" would not be held against an individual in seeking to access or retain a diagnosis. In this, we were borrowing a formulation that had been very successful in the Americans with Disabilities

Act Amendments Act of 2008, legislation ASAN had successfully advocated for ensuring that individuals would not lose the legal protections of the ADA if they successfully used "mitigating measures" to manage their disability.

Strategy and Tactics

In order to advance the priorities and protect against the concerns previously discussed, ASAN pursued a combination of social, political, and scientific strategies to "lobby" the DSM-5 process. Ultimately, our work was rooted in a simple reality, often obscured given the inscrutable nature of the process of making the DSM: it was written by people, and people can be communicated with, influenced, and convinced, even when they are autism researchers.

Early on, we made a judgment call that the autistic community, though possessed with an (in our opinion) indisputable moral claim to be represented in the DSM-5 process on an equal basis, lacked any material leverage with which to pressure the APA to include us on a formal basis or to accede to demands regarding modifications to the criteria. By this time, ASAN leadership had become experienced in running grassroots campaigns designed to secure autistic community priorities, even against opposition. We regularly conducted what would be referred to as a "pressure points" analysis in the leadership training we would later run for autistic college student organizers: identifying the levers through which advocacy could influence a target into complying with the autistic community's demands or making concessions toward those ends.

In the case of the APA, no material "pressure points" presented themselves. As such, even though ASAN was perceived as a more "militant" organization vis a vis the autism research and clinical worlds, Ari made a decision to operate a campaign based primarily on personality, persuasion, and evidence from the research literature. Our philosophy was always (and remains) using whatever tool is most effective for a particular job. Thus, a decision was made to cultivate relationships with individual workgroup members and the workgroup as a whole with the goal of convincing them to advocate for our priorities and to provide advance copies of working drafts.

While Ari and Steven were the organization's primary leads on DSM-5 advocacy, others played critical roles. Scott Robertson, ASAN's co-founder and then a member of ASAN's board and a PhD candidate, also assisted the production of early documents sent to the workgroup and participated in early phone calls, as did Paula Durbin-Westby, an autistic activist on the board of ASAN and later to join the organization's staff. Zoe Gross, then an intern with ASAN and later to become the organization's Director of Operations, drafted critical background material provided to the workgroup on the challenges facing autistic women and autistic people of color in accessing a diagnosis and the resulting disparities these groups faced. She also provided illustrative examples regarding circumstances under which individuals might fall out of the boundaries of early drafts of the criteria, while still needing the support and recognition that an autism diagnosis could provide. Amanda Vivian, an autistic writer and creator of the Autistic Passing Project (http://autisticpassing.tumblr.com/), provided critical feedback on early drafts of ASAN feedback, among others.

While Steven provided research knowledge and scientific analysis to ASAN's work on influencing the DSM-5 continuously throughout the organization's advocacy, this intensified after he signed a contract in 2011. He led the writing of most memos and authored several independently. Topics included documenting the social abilities and social interest and empathy of autistic people, motor and movement issues, differential diagnosis, gender and race disparities, addressing potential misunderstandings of autistic activists and the neurodiversity movement, diagnostic practice, considerations for why the revision might "miss" autistic people, and so on. For specific and sensitive matters, he sometimes communicated directly with Members B or C (see below).

Communications with the Workgroup

In 2009, ASAN made contact with the DSM-5 Workgroup through one of its members, hereby referred to as Member A, whom Ari had corresponded with earlier regarding early intervention methodology. The two had earlier found common ground over a shared critique of the excess rigidity of behaviorist interventions. Separately, Ari connected with the workgroup Chair at a meeting of the Interagency Autism Coordinating Committee (IACC) and, after Member A provided the Chair and Workgroup with a favorable impression of ASAN, Ari was invited to provide written and verbal feedback to the workgroup at several teleconferences and semi-annual in-person meetings in Washington DC hotel rooms. Ari also used the IACC as a vehicle for highlighting autistic community priorities and concerns regarding the DSM-5 during his two years as a public member of the committee (for more details on this from the perspective of an external observer at the IACC meetings, see Moore, pp. 169–198 [19]).

After an individual meeting with the Chair and phone calls with her and Member A, Ari met with the workgroup in person on the morning of April 8, 2010 (a meeting for which Steven provided significant research support). At this meeting, Ari stressed the importance of acknowledging "mitigating measures" and ensuring that individuals would not lose access to a diagnosis by virtue of their having learnt how to "pass" as non-autistic, a serious concern for many autistic adolescents and adults.

In addition, Ari stressed ASAN's opposition and concern regarding the severity scale, both in general and in its current formulation. ASAN was (and remains) worried that the introduction of a severity scale would be used by clinicians and service providers to set inappropriate "service goals" focusing on making autistic children and adults look and act "less autistic". We were particularly concerned by the fact that, at the time, drafts of the severity scale included references to "fixated interests", suggesting that clinicians and other professionals should try and redirect autistic children away from their passionate special interests, and to "repetitive motor movements", which many autistic people enjoy and which help us to self-regulate (and which we reclaim as *stimming*). Ari also indicated ASAN's concern with the draft criteria's emphasis on "social reciprocity", a vague concept whose most common clinical measures ASAN considered to be flawed.

This meeting was well-received by the workgroup, leading to a growing correspondence between ASAN and workgroup members both individually and collectively. ASAN soon made contact with Member B and Member C, who along with the Chair and Member A corresponded with Ari and Steven to help inform the workgroup's deliberations. These Members did not necessarily agree with all of ASAN's recommendations—indeed, it was common for us to work with members on one set of priorities (i.e.: opposition to the severity scale, maintaining a sufficiently broad diagnosis, etc.) who disagreed with us on another set of priorities (i.e.: opposition to recovery criteria, etc.).

This made the establishment of relationships with as many members of the workgroup as possible a high priority. Ari, Steven, and others went to conferences in the US, Canada, and the UK—the home countries of workgroup members—where we knew that members would be present in order to make contact, establish a social relationship, and parlay that into communicating our recommendations and collecting intelligence on the current status of the draft criteria. At times, this resulted in drafts being provided to our team from individual workgroup members, to which ASAN provided specific and substantive comments (with academic references as appropriate). Other times, it simply resulted in the collection of useful observations on the attitudes of individual workgroup members toward our recommendations and their recollections of internal deliberations within the workgroup.

ASAN continued to correspond and meet with the DSM-5 workgroup members, though usually we were not invited to participate directly in workgroup calls and meetings, with a few exceptions. For example, ahead of the November 2011 meeting, ASAN developed a private memo making recommendations on the latest draft of the criteria, in particular urging a revision of the social communication domain from requiring 3 of 3 sub-criteria to qualify for a diagnosis to only requiring 2 of 3 criteria. (We also proposed an alternative recommendation of adding a fourth regarding language and speech issues, to require 3 of 4). This recommendation was not accepted, though others reflected in the memo were.

We also pushed for acknowledgment of motor movement issues and for strengthening of the language acknowledging that different contexts informed whether or not autistic traits would be visible. At the time, this language stated only that:

> symptoms must be present in the early developmental period (but may not become fully manifest until social demands exceed limited capacities).

We recommended that the language be revised to read as follows:

> Symptoms must be present in early childhood (but may not become fully manifest until social demands exceed limited capacities, or because of compensatory or coping mechanisms developed over time).

The final criteria closely followed this formulation, reading:

> Symptoms must be present in early childhood (but may not become fully manifest until social demands exceed limited capacities, or may be masked by learned strategies in later life).

The memo also urged the inclusion of greater material on adults, women and girls, racial and cultural minorities, socioeconomic status and other factors that influenced disparities in access to diagnosis, in the accompanying text, and provided the workgroup with illustrative examples for each. Finally, we urged the elimination of the severity scale and provided guidance for the accompanying text on differential diagnosis.

On the sidelines of the meeting, Ari met with Member B and Member C, communicating with each individually during breaks and the lunch period. This correspondence from Ari to Steven and Zoe, redacted to avoid disclosing the names of the workgroup members, provides some insight into the nature of these interactions:

> A few highlights, while they're fresh. Needless to say, none of this is for repetition or forwarding under any circumstances:
>
> - Met with [Member B] and [Member C] for 20 min., they report our document was well received by the Committee. I snuck a peek into their folders when I got there: every member of the ND Work Group had received a copy of our memo. [Member B]'s looked like it had been leafed through decently and they say they made use of it throughout the morning. Good job, team!:) Your hard work was not for naught.
> - They backed our severity scale concerns, said the dsm v apa folks requiring it of everyone, said they'd be willing to put language in accompanying text clarifying it not intended as proxy of treatment goals and outcomes, shouldn't be used as measure of service provision need. Pushed a bit more, they said they were open to dropping fixated interests

(and maybe rrbs?) from the scale and using a flexibility/ef measure along with social comm. instead. I pointed out that this might lead to more work occurring around self regulation.
- Focused mostly on accompanying text and severity, as they clarified that while the criteria MIGHT get opened up later, they've been instructed to leave it for now until field trial data comes back;
- Some willingness to elaborate on motor and language issues in accompanying text, said it was already there to some degree, they might expand on it;
- Our first discussion focused on how to capture adults in diagnosis who were hard to ID. They asked two starting questions: "what services did this population need and how would we suggest they guide a junior clinician who hasn't seen asd before as to how to identify these individuals?" Strongly emphasized that even those who don't require traditional types of service provision might still benefit from diagnosis to access ADA protections, reasonable accommodation and support groups. ([Member C] had tried to raise concern on "political motivations to access diagnosis" but this helped mitigate that concern or at least convince [Member C] that wasn't our motivation). Also pointed out that accurate diagnosis useful for clinicians providing treatment for co-occurring mh conditions like anxiety and depression.

We had a discussion on coping mechanisms (they referred to this as "scaffolding" and "masking") and the risk of individuals losing their diagnosis or not getting one in the first place. This was where we had more disagreement. [Member C] feels strongly that there are large numbers of people seeking an asd diagnosis who "just don't meet the criteria" as a way of escaping "legal, workplace or marital" problems. We pushed back here.

Discussed mechanisms of addressing masking in diagnostic process, I suggested greater weight to self report, [Member C] disagreed, citing again the supposed fakers trying to get asd diagnosis that doesn't fit. Respectfully disagreed, then reinforced that "do no harm" principle means that its better we capture a few folks that don't fit than risk pushing off folks who do. Not much agreement there.

Moved onto other potential ways for assessing what we both referred to as "cognitive impacts" for those who effectively "mask" behavioral traits. They were very interested in using anxiety and depression as possible proxies to catch those who are experiencing cognitive impacts due to masking. Pushed for inclusion in accompanying text.

- Steven, [Member C] disagrees strongly with your assessment re women and girls having distinct traits, feels there are "hundreds of studies that disagree with the few you cited". I pushed back here too, stated that while lit for bio based differences may be unclear, there is strong lit often outside of asd field for differences in manifestations between boys and girls due to upbringing, social context, etc. Pushed for acknowledgement in accompanying text.
- They agreed that a "subclinical" category on the severity scale was, on further consideration, a bad idea. They said it was intended to capture those who felt, "they had a condition, not a disorder" ([Member C] again). I stated that this is likely a corruption of Neurodiversity philosophy, these folks were trying to say "disability, not disease". We agreed - particularly [Member B] who was consistently more friendly - that there was real risk that a subclinical category could push folks "off the spectrum". I pointed out that it is unlikely those with a "subclinical severity" could access ADA and 504 protections.
- Whenever possible, I tried to move conversation to legal/policy impacts of their decisions, they don't understand law & policy and know we do, thus they're more likely to hear from us on those points. Made it very explicit throughout we had no intention of making "political/identity" arguments, only "practical/research and policy driven ones". They appreciated that."

As reflected in the above report, one of many written by Ari and Steven in their respective interactions with the workgroup or its individual members, ASAN had a complex relationship with the individuals we communicated with on the workgroup, some of whom shared most of our views while others agreed with us on only a few things. Some possessed views that we found extremely objectionable, requiring careful calibration in our communications with them to preserve the relationship while pushing back on viewpoints that had the potential to deeply harm our community if they were incorporated into the DSM-5 criteria.

Because of the power imbalance between the APA and the autistic community, and the tremendous impact that the DSM 5 could have on our community, we felt that an "inside game" was the most effective way we could promote change, thus our willingness to de-emphasize "political/identity" arguments. There is, of course, a certain irony here, in that the "legal/policy impacts" of the DSM-5 are unquestionably political, but as those with decision-making authority in the process tended to present themselves as engaged in an apolitical endeavor, we adjusted our rhetoric accordingly to maximize effectiveness.

On January 31, 2012, ASAN issued a joint statement on the DSM-5 with the Autism Society of America, a parent-led group that we had an uneasy détente with, urging the "DSM-5 Neurodevelopmental Disorders Working Group to interpret the definition of autism spectrum disorder broadly, so as to ensure that all of those who can benefit from an ASD diagnosis have the ability to do so" [20]. This was made possible both by early efforts to build up a relationship between ASAN and what we then perceived as more moderate elements within the autism parent movement and the fact that concerns over the DSM-5 extended across traditional dividing lines of self-advocate and parent perspectives in autism. These concerns grew in response to a headlined *New York Times* report published days earlier about a preliminary study by the former chair of the DSM-IV workgroup, which warned that about 75% of people diagnosed with Asperger's and 85% of people diagnosed with PDD-NOS would no longer be eligible for an autism spectrum diagnosis [7]. While the Autism Society was a larger and better-funded organization, they had not built up significant internal technical expertise on the legal, policy, or research questions at issue within the DSM-5, requiring them to rely on our expertise as their concerns grew.

In June 2012, Ari and Steven released two policy briefs, timed to coincide with the final public comment period on the DSM-5, for which ASAN issued talking points to our grassroots in May [21]. The first, entitled "What Are the Stakes? An Analysis of the Impact of the DSM-5 Draft Autism Criteria on Law, Policy and Service Provision" provided comprehensive analysis of the implications of DSM-5 proposals on legal, policy, and service-provision systems. In this policy brief, we presented distinctive analysis that in special education, non-discrimination protections and rights to reasonable accommodations, developmental disability services, and income support, a shift to a single unified diagnosis would likely increase access to publicly funded service provision [22]. We also called attention to the fact that the proposed non-autism diagnosis of Social Communication Disorder, created by the workgroup in part to house those who might be pushed off the autism spectrum, would likely be less useful in assisting individuals to gain access to services [22]. Later, a Workgroup member cited both policy briefs, referring to this first one as one of only three papers "of major importance" published on the then-pending criteria [23].

The second policy brief, entitled "ASD in DSM-5: What the Research Shows and Recommendations for Change" provided an academic evidence base for our concerns and specified our recommendations. The policy brief analyzed the draft criteria's likely impact on under-represented groups, placing particular emphasis on adults, women and girls, and racial and ethnic minority groups, and made another case for acknowledging motor/movement difficulties within the criteria. We also made several technical edits, and recommendations to address concerns of the revision pushing individuals off the autism spectrum (particularly due to the uniquely stringent social communication requirement). For example, we recommended attaching the Social Communication Disorder diagnosis to the autism spectrum, "possibly by renaming it as ASD-Not Elsewhere Classified or ASD-Social Communication subtype", increasing its utility as a means of accessing services. The policy brief was deeply grounded in the research literature, with 216 different citations of a wide array of peer-reviewed autism research studies [16]. A Workgroup member cited it within a study applying the DSM-5 criteria to adults, agreeing based on their own research that the minimum requirements for meeting criteria could be relaxed to correctly identify more people as autistic without significantly adding false positives [24].

Our final engagement with the WorkGroup took place at an in-person meeting in late 2012, when Ari was invited to attend the last meeting of the DSM-5 Neurodevelopmental Disorders WorkGroup before the criteria were finalized. There, he reiterated our concerns regarding sensitivity and made a final impassioned plea to consider loosening the social communication domain or linking Social Communication Disorder to the spectrum. Michael John Carley of GRASP also received an opportunity to comment via phone, reiterating GRASP's opposition to the loss of the separate Asperger's diagnosis. Though we did not succeed in achieving all of our goals, we nonetheless substantially influenced the final diagnostic criteria and the accompanying text.

Outcomes and Implications

ASAN's effort to lobby the DSM-5 is historically significant in that it represents the first successful effort of the autistic community—and as far as we are aware, any disability community—to successfully influence the modification of their own diagnostic criteria. While communities have successfully advocated to eliminate a diagnosis from the DSM (i.e.: homosexuality) or to incorporate one, we are aware of no prior example of successful advocacy to refine and improve diagnostic criteria from the community subject to it.

Having said that, we were only partially successful at achieving our advocacy objectives, owing in large part to the lack of any formal recognition of the value of autistic input in the development of the criteria earlier in the process. While the Neurodevelopmental Disorders Workgroup ultimately chose to acknowledge Ari as a formal advisor to the DSM-5, it did so only after the criteria had been finalized. Even then, they did not inform us ahead of the fact that this was planned. (Had they done so we would have pushed to formally acknowledge Steven's role as well.) Nevertheless, the Workgroup Chair singled ASAN out for praise before international researchers [25], as did another member before the autism community, thanking us for our "steadfastness in tracking diagnostic criteria", which he said had been "extraordinarily helpful" [26].

And yet, the vast majority of workgroup meetings took place without autistic input, with only a small number of direct contacts between ASAN personnel and the workgroup as a whole (as distinct from the successful cultivation of some individual members). Autistic input in the DSM-5 ultimately took the form of an intelligence operation, requiring the licit and illicit cultivation of assets to collect partial information on potential revisions and inform formal communications regarding requested changes. It would have been far preferable for the autistic community to have received a direct and acknowledged seat at the table.

Perhaps because of this lack of formal recognition, only some of ASAN's goals were accepted into the DSM-5. The unique diagnostic needs of adults (including that allowing self-report may strengthen the assessment process, particularly for those who lack relatives with access to their clinical history), women and girls, and racial and ethnic minority groups were incorporated into the accompanying text, though not in as much detail as we had pushed for.

No "recovery" criteria were incorporated into the diagnosis, and the severity scale includes no reference to "subclinical" autism, a category that would have made it substantially harder for those included within it to access legal protections and service provision. Both are likely the result of our efforts. In addition, modifications were made to the severity scale that mitigated its potential harms, though it was still incorporated against our recommendation and even the Workgroup's objection [26].

The APA required a severity scale as part of all diagnoses in the DSM-5, yet ASAN's influence led to several notable concessions regarding it and related text. The Workgroup reframed the scale as about need for support because individuals might function well *because* of support and we did not want that support taken away. As Workgroup Chair Sue said, "I think the example that was given to us [by Ari], if you need a crutch to be able to walk, but you walk perfectly fine with that crutch, you don't want to, then, say you don't need that crutch anymore" (p. 198) [19]. Similarly, after rejecting our call to eliminate the scale altogether due to APA's insistence, the Workgroup adopted ASAN's backup recommendation to try to defang the scale by prohibiting its use for services: the accompanying text to the DSM-5 states that "the descriptive severity categories should not be used to determine eligibility for and provision of services" and that "these can only be developed at an individual level and through discussion of personal priorities and targets" (p. 51) [1]. Furthermore, the emphasis on inflexibility or executive functioning in, and removal of "fixated interests" from, the restrictive and repetitive behaviors domain of the scale, result from ASAN's involvement. In parallel, the accompanying text states, "Special interests may be a source of pleasure and motivation and provide avenues for education and employment" (p. 54) [1].

In other respects too, the final text reflected ASAN's argument that autistic people's manifestations of their autism and functioning vary too much to be applied systematically to service provision and clinical practice. The main text's clarification that the examples given "are illustrative, not exhaustive" closely followed ASAN's recommendation [16]. The Workgroup adopted our recommendation to loosen the requirement for social communication deficits "across contexts" to "in multiple contexts" in the main text, and limited the emphasis on relationship deficits to those with *peers*. Similarly, it noted the context-dependent nature of autistic people's functioning multiple times in the accompanying text. Through our comments on confidential drafts of the diagnostic text, ASAN successfully encouraged language recommending that multiple sources of information be used together in assessment to identify behaviors that do not always present clinically, such as direct observation and interaction, interview on history, and other reports, which can dramatically increase the likelihood of identifying autism [27, 28]. As concessions to our input, the Workgroup added language noting uneven skills and a common gap between IQ and lower adaptive behavior—which challenges the notion of "high-functioning" autism. It likewise added advice that autistics with limited language may show strengths on nonverbal, untimed cognitive tests—which challenges "low-functioning" or "severe" autism tropes.

Perhaps most importantly, the inclusion of ASAN's requested language allowing diagnosis "currently or by history" as well as acknowledging that "symptoms…may be masked by learned strategies in later life" (described further as effortful and taxing in the accompanying text) offers meaningful opportunities for autistic adults to be diagnosed at greater rates than they have been previously. The inclusion of this language likely mitigated some of the anticipated narrowing of the diagnosis and opened up opportunities to address diagnostic disparities, especially on the basis of age. Other quieter influences more literally ensured access to diagnosis, such as retaining the ability to diagnose OCD alongside autism (we provided ample studies differentiating them).

Nonetheless, evidence does suggest that some narrowing did take place. Studies applied prospectively that compare DSM-IV with DSM-5 criteria reported that the latest revision narrowed eligibility for an autism spectrum diagnosis by between 4% [29] and more than 10% among children, with higher proportions missed for children with previous Asperger's (20%) and especially PDD-NOS (75%) diagnoses [30]. The DSM-5 particularly missed girls, older children, and children with subtler autistic behaviors [30]. Still, preliminary evidence does suggest DSM-5 increases access to services (e.g. in special education) for those diagnosed [29], and likely the revision would have missed many more people were it not for ASAN's efforts.

Had our recommendation to require only two of three criteria in the social communication domain been accepted, this narrowing would likely not have taken place, or would have not taken place to the same degree. The DSM-5 will likely "miss" more people as individuals increasingly get assessed for the autism spectrum for the first time. Those who already had a diagnosis have a limited amount of protection, as the Workgroup ceded to pressure by seeking to soften the transition to the new system through the following language: "Individuals with a well-established DSM-IV diagnosis of autistic disorder, Asperger's disorder, or pervasive developmental disorder not otherwise specified should be given the diagnosis of autism spectrum disorder" (p. 51) [1].

Emerging evidence on the implementation of DSM-5 confirms our concern about the severity scale. It is our experience that when categories of "severity" are provided they inevitably are used as mechanisms for ascertaining service eligibility or service/treatment goals. Indeed, the National Insurance Disability Scheme in Australia, enacted since the completion of the DSM-5, interprets the lowest support level ("Requiring support") as usually disqualifying autistic people from eligibility for services (https://www.ndis.gov.au/). Fortunately, academic studies suggest the "need for support" framing has shifted some attention toward disability (low adaptive functioning) as well as non-specific disability characteristics sometimes associated with autism such as low cognitive ability [31, 32], without a consistent relationship between these developmental domains and core autism "symptoms" [33].

This is a good thing, as we have always believed that the focus of service provision should be on improving adaptive functioning and other person-centered goals, rather than trying to "correct" or "cure" autistic traits. If the scale is being used as a guide for service or treatment goals, then it is particularly fortunate that ASAN secured the removal of "fixated interests" and "repetitive motor movements" as measures within the severity scale, given the importance of "special interests" and "stimming" to many autistic people and the clear autistic preference for services oriented toward improving happiness and quality of life rather than the enforced imitation of "typical" behavior and appearance.

In contrast, the introduction of the social communication disorder (SCD) diagnosis appears less damaging because it is rarely utilized. As a major communication scholar and autism researcher put it, "Entry into the DSM...has not changed anything: There are no new assessment tools, no clear diagnostic criteria, no stronger evidence for the existence of the condition and no innovative, effective interventions" [34]. It has attracted little interest in practice: "Whatever the reason, most expert clinicians do not find the new diagnosis necessary or useful", she added. Nor do researchers, as "more than 10,000 papers have the term 'autism' in the title" compared with "just 10 papers on 'social communication disorder'".

Neurodiversity activists deserve some credit for the dearth of diagnoses of SCD, as we have helped to improve attitudes toward autism such that the SCD diagnosis rarely gets assigned to reduce stigma (as the Asperger's diagnosis once was used). This apparently almost unused diagnosis further validates ASAN's approach to not let the supposedly greater stigma of "autism" interfere with a unified spectrum diagnosis. The relatively low utilization of the diagnosis is positive, given our longstanding concern that a SCD diagnosis would open up access to significantly less support than an autism diagnosis does. Nonetheless, we continue to believe that the social communication domain of the autism criteria should be loosened. Indeed, a large study using major databases found that more than four times as many autistic children failed to meet the social communication domain requirement only (more than 6.2%) as compared to the restricted and repetitive behavior domain (less than 1.5%; Huerta et al. [35]). We activists emphasize that social communication always results from broad factors within and between people [36, 37].

These kinds of "practical" knowledge of the other side of the service system support our later recommendation that future iterations of the DSM should formally include autistic input on the workgroup.

Moving forward, we make the following recommendations for future consideration:

1. **Acknowledge the DSM as a Political Process *and* a Scientific One:** While ASAN was careful to root our advocacy regarding the DSM in scientific rather than political language, we always understood the process as both a political and a scientific one. We mean this in a non-pejorative way, simply acknowledging the reality that any effort to articulate a diagnostic criteria will have distributive consequences in

terms of public resources and social consequences in terms of identity. In the future, we urge others to acknowledge the political dimension of the DSM, not with the intent of denigrating the process, but to allow open consideration of factors that influence those writing the criteria and to acknowledge the consequences of those criteria on those that are subject to it. Similarly, we urge other communities to learn from our example in exploring how they too can play a larger role in influencing their diagnoses, while leveraging scientific knowledge.

2. **Provide for Autistic Representation in Future Revisions**: The next iteration of the DSM should provide an opportunity for autistic voices to be represented as full and equal partners within the workgroup developing the criteria. This reflects both the considerable sophistication of the autistic advocacy community in understanding and engaging with the research literature and the moral claim for representation, consistent with the longstanding disability rights principle of "Nothing About Us, Without Us!" Such representation may take multiple forms, both involving organized autistic-run groups like ASAN, and the growing number of openly autistic researchers with expertise in autism, an increasingly common phenomenon.

3. **Abandon the new Severity Scale and the SCD Diagnosis**: The political dimension was not lost on all of the Workgroup—one member acknowledged that the group introduced the SCD diagnosis for "political and health reasons…DSM-5 was not a scientific process…the empirical evidence is *not* in support of social pragmatic disorder" (Lord in [38]. Furthermore, the inconsistent relationship between the degree of core autistic traits, intellectual ability, and adaptive functioning, as well as lack of consensus on how to measure "severity" in autism [39], suggest the need to abandon the severity scale. If they are retained, further research is also needed on the service and clinical implications of both the severity scale and the frequency of the SCD diagnosis. It is our belief that such research would ultimately validate our view that they should be eliminated.

4. **Prioritize Research into the Distributive Implications of the DSM-5 across Groups**: Preliminary evidence suggests that much work remains to close racial and gender disparities in access to diagnosis, and that class, age and geography remain as factors in who gets access

to a diagnosis and who does not [29]. Research should be conducted to ascertain if DSM-5 has led to a narrowing or expansion of the autism diagnosis, identifying which groups have been impacted in which way, and understanding the particular aspects of the new criteria that are contributing to that outcome. Particular priority should be given to understanding the impact of the criteria on adults and autistic people with less obvious traits.

5. **Allow for Near-Term Further Revisions**: As new information becomes available on the implications of the DSM-5 criteria, the APA should acknowledge the need for a DSM-5.1, 5.2, 5.3, etc. before the DSM-6 process begins. Because so much information used by clinicians is now provided online rather than in textbooks, APA has an unprecedented opportunity to deliver revisions to the criteria on a faster timetable than in the past, while still acknowledging the importance of stability and careful deliberation. This revision process should include autistic voices as full partners and prioritize ensuring that autistic people are not adversely impacted by loss of access to the diagnosis and resulting services and legal protections.

The DSM is not provided on stone tablets brought down from a mountain—it is a document, written by people, and as such can be influenced using creativity, evidence, and strategic argument. Historically, critiques by disabled people of the DSM have often been critiques of psychiatry itself, either in general or in terms of its specific applicability to particular groups. This has not lent itself to collaboration between clinicians and disabled activists, since the latter tend to see the DSM itself as illegitimate. Some associated with the "anti-psychiatry" movement even reject the idea that diagnoses represent actual underlying neurological differences from the norm as opposed to purely responses to trauma.

ASAN's perspective is different and is instead rooted in the idea of "neurodiversity", which challenges the "medical model" that assumes that the goal of service provision or "treatment" is to restore autistic people to "normalcy" or, as Lovaas put it, indistinguishability from peers [40]. While we reject the idea that interventions should stress "indistinguishability" and often challenge the idea of exclusive medical authority, we do not reject the utility of the autism diagnosis itself or the well-documented reality that it constitutes a real divergence from "typical" neurology.

In short, we largely agree with psychiatry as to what autism is (a difference of neurology) and feel that scientific research should play a key role in defining the diagnosis. Nonetheless, we believe that identification of autism should transition to a non-pathological system that allows inclusion of evidence-based neutral differences and strengths, recognizing that autistic traits can be strengths, challenges, or neutral depending on context (and are often deeply valued by autistic people ourselves; Russell et al. [41]). And while we agree with psychiatry that autism emerges from neurological differences, we disagree with many assessments of how autistic people should be treated—and wish to call attention to the social, legal, and political context in which research and diagnosis take place.

The neurodiversity movement, as we understand it, is in creative tension with mainstream psychiatry, not in opposition [42]. This is useful in that we are able to articulate an important critique as to how autistic people are treated while agreeing that the autism diagnosis delivers value and should be maintained. This is the theoretical framework that allowed our collaborative approach to DSM-5 advocacy to be as successful as it was.

To complete that success, however, psychiatry must acknowledge the autistic community (and other similar communities) as an equal, not as a junior partner. Future iterations of the DSM should include autistic people within the process in an explicit and acknowledged fashion, sparing us and them the aggravations, inefficiencies, and hypocrisies inherent in our needing to launch complex influence operations to have our views represented. It is our sincere hope that as the worlds of autism research and clinical practice continue to mature, such a partnership will take form.

As always, Nothing About Us, Without Us!

References

1. American Psychiatric Association. (2013). *Diagnostic and Statistical Manual of mental disorders (DSM-5®)*. Washington, DC: Author.
2. Kaufmann, W. E. (2012). *The new diagnostic criteria for autism spectrum disorder*. Autism Consortium. Retrieved from https://www.youtube.com/watch?v=in23jSkL1eA&fbclid=IwAR1ngCRHryNMOgzOeOVtK_FewIHwF6Tvntbi0EXbcY3xL7KmfCne7TMGiMc.
3. Lord, C. (2012). *Rethinking autism diagnoses*. WCARP Autism Symposium. Retrieved from https://www.youtube.com/watch?v=LX6rRWibX4E.
4. American Psychiatric Association. (1994). *Diagnostic and Statistical Manual of mental disorders: DSM-IV*. Washington, DC: Author.
5. Harmon, A. (2012, January 20). A specialists' debate on autism has many worried observers. *The New York Times*. Retrieved from https://www.nytimes.com/2012/01/21/us/as-specialists-debate-autism-some-parents-watch-closely.html.
6. Jabr, F. (2012, January 30). *By the numbers: Autism is not a math problem*. Scientific American. Retrieved from https://www.scientificamerican.com/article/autism-math-problem/.
7. Carey, B. (2012, January 19). New definition of autism will exclude many, study suggests. *New York Times*. Retrieved from https://www.nytimes.com/2012/01/20/health/research/new-autism-definition-would-exclude-many-study-suggests.html.
8. Swedo, S. E., Baird, G., Cook, E. H., Happé, F. G., Harris, J. C., Kaufmann, W. E., et al. (2012). Commentary from the DSM-5 workgroup on neurodevelopmental disorders. *Journal of the American Academy of Child & Adolescent Psychiatry, 51*(4), 347–349.
9. Giles, D. C. (2014). 'DSM-V is taking away our identity': The reaction of the online community to the proposed changes in the diagnosis of Asperger's disorder. *Health, 18*(2), 179–195.
10. Lord, C., Petkova, E., Hus, V., Gan, W., Lu, F., Martin, D. M., & et al. (2012). A multisite study of the clinical diagnosis of different autism spectrum disorders. *Archives of General Psychiatry, 69*(3), 306–313.
11. American Psychiatric Association. (2011). Proposed revision: Autism Spectrum Disorder. Retrieved January 25, 2011 from http://www.dsm5.org/ProposedRevision/Pages/proposedrevision.aspx?rid=94#. Retrieved from http://www.thinkingautismguide.com/2012/01/dsm-5-autism-criteria-clarifying-impact.html.

12. Hamilton, J. (2010). *Asperger's officially placed inside autism spectrum.* National Public Radio. Retrieved from https://www.npr.org/templates/story/story.php?storyId=123527833.
13. Carley, M. J. (2012). *Don't reduce the criteria for an autism spectrum condition in the DSM-5.* Change.org. Retrieved from https://www.change.org/p/dsm-5-committee-dont-reduce-the-criteria-for-an-autism-spectrum-condition-in-the-dsm-5.
14. Broderick, A. A., & Ne'eman, A. (2008). Autism as metaphor: Narrative and counter-narrative. *International Journal of Inclusive Education, 12*(5–6), 459–476.
15. Mattila, M. L., Kielinen, M., Linna, S. L., Jussila, K., Ebeling, H., Bloigu, R., et al. (2011). Autism spectrum disorders according to DSM-IV-TR and comparison with DSM-5 draft criteria: An epidemiological study. *Journal of the American Academy of Child and Adolescent Psychiatry, 50*(6), 583–592.
16. Kapp, S., & Ne'eman, A. (2012). *ASD in DSM-5: What the research shows and recommendations for change.* Washington, DC: Autistic Self Advocacy Network.
17. Mukaddes, N. M., Mutluer, T., Ayik, B., & Umut, A. (2017). What happens to children who move off the autism spectrum? Clinical follow-up study. *Pediatrics International, 59*(4), 416–421.
18. Olsson, M. B., Westerlund, J., Lundström, S., Giacobini, M., Fernell, E., & Gillberg, C. (2015). "Recovery" from the diagnosis of autism—And then? *Neuropsychiatric Disease and Treatment, 11,* 999.
19. Moore, M. J. (2014). *On the spectrum: Autistics, functioning, and care* (Doctoral dissertation, UC Santa Cruz).
20. Autistic Self Advocacy Network (ASAN). (2012, January 31). *Joint ASAN-Autism Society statement on DSM-5.* Retrieved from https://autisticadvocacy.org/2012/01/joint-asan-autism-society-statement-on-dsm-5/.
21. Autistic Self Advocacy Network (ASAN). (2012, May 7). *ASAN Talking points on DSM-5.* Retrieved from https://autisticadvocacy.org/2012/05/asan-talking-points-for-public-comment-on-dsm-5-autism-spectrum-disorder-criteria/.
22. Ne'eman, A., & Kapp, S. (2012). *What are the stakes? An analysis of the impact of the DSM-5 draft autism criteria on law, policy and service provision.* Washington, DC: Autistic Self Advocacy Network.
23. Mahjouri, S., & Lord, C. E. (2012). What the DSM-5 portends for research, diagnosis, and treatment of autism spectrum disorders. *Current Psychiatry Reports, 14*(6), 739–747.

24. Wilson, C. E., Gillan, N., Spain, D., Robertson, D., Roberts, G., Murphy, C. M., et al. (2013). Comparison of ICD-10R, DSM-IV-TR and DSM-5 in an adult autism spectrum disorder diagnostic clinic. *Journal of Autism and Developmental Disorders, 43*(11), 2515–2525.
25. Swedo (S. E.) (and members of the DSM-5 Neurodevelopmental Disorders Workgroup). (2012, May 18). *An update on the DSM-5 recommendations for autism spectrum disorder and other neurodevelopmental disorders.* International Meeting for Autism Research, Toronto, Canada.
26. King, B. (2012). *Keynote presentation.* Annual meeting of the Autism Society of America, San Diego, CA.
27. Foley-Nicpon, M., Fosenburg, S. L., Wurster, K. G., & Assouline, S. G. (2017). Identifying high ability children with DSM-5 autism spectrum or social communication disorder: Performance on autism diagnostic instruments. *Journal of Autism and Developmental Disorders, 47*(2), 460–471.
28. Mazefsky, C. A., McPartland, J. C., Gastgeb, H. Z., & Minshew, N. J. (2013). Brief report: Comparability of DSM-IV and DSM-5 ASD research samples. *Journal of Autism and Developmental Disorders, 43*(5), 1236–1242.
29. Baio, J., Wiggins, L., Christensen, D. L., Maenner, M. J., Daniels, J., Warren, Z., et al. (2018). Prevalence of autism spectrum disorder among children aged 8 years—Autism and Developmental Disabilities Monitoring Network, 11 Sites, United States, 2014. *MMWR Surveillance Summaries, 67*(6), 1–23.
30. Mazurek, M. O., Lu, F., Symecko, H., Butter, E., Bing, N. M., Hundley, R. J., & et al. (2017). A prospective study of the concordance of DSM-IV and DSM-5 diagnostic criteria for autism spectrum disorder. *Journal of Autism and Developmental Disorders, 47*(9), 2783–2794.
31. Mazurek, M. O., Lu, F., Macklin, E. A., & Handen, B. L. (2018). Factors associated with DSM-5 severity level ratings for autism spectrum disorder. *Autism.* https://doi.org/10.1177/1362361318755318.
32. Mehling, M. H., & Tassé, M. J. (2016). Severity of autism spectrum disorders: Current conceptualization, and transition to DSM-5. *Journal of Autism and Developmental Disorders, 46*(6), 2000–2016.
33. Gardner, L. M., Campbell, J. M., Keisling, B., & Murphy, L. (2018). Correlates of DSM-5 autism spectrum disorder levels of support ratings in a clinical sample. *Journal of Autism and Developmental Disorders, 48*(10), 3513–3523.

34. Tager-Flusberg, H. (2018). Why no one needs a diagnosis of 'social communication disorder'. *Spectrum*. Retrieved from https://www.spectrumnews.org/opinion/viewpoint/no-one-needs-diagnosis-social-communication-disorder/.
35. Huerta, M., Bishop, S. L., Duncan, A., Hus, V., & Lord, C. (2012). Application of DSM-5 criteria for autism spectrum disorder to three samples of children with DSM-IV diagnoses of pervasive developmental disorders. *American Journal of Psychiatry, 169*(10), 1056–1064.
36. Donnellan, A. M., Hill, D. A., & Leary, M. R. (2010). Rethinking autism: Implications of sensory and movement differences. *Disability Studies Quarterly, 30*(1). Retrieved from http://dsq-sds.org/article/view/1060/1225.
37. Kapp, S. K. (2013). Empathizing with sensory and movement differences: Moving toward sensitive understanding of autism. *Frontiers in Integrative Neuroscience, 7,* 38.
38. Spectrum. (2013, May 29). *DSM-5 live discussion*. Retrieved from https://www.spectrumnews.org/features/special-report/live-dsm-5-discussion/.
39. Weitlauf, A. S., Gotham, K. O., Vehorn, A. C., & Warren, Z. E. (2014). Brief report: DSM-5 "levels" of support:" a comment on discrepant conceptualizations of severity in ASD. *Journal of Autism and Developmental Disorders, 44*(2), 471–476.
40. Lovaas, O. I. (1987). Behavioral treatment and normal intellectual and educational functioning in autistic children. *Journal of Consulting and Clinical Psychology, 55,* 3–9.
41. Russell, G., Kapp, S. K., Elliott, D., Elphick, C., Gwernan-Jones, R., & Owens, C. (2019). Mapping the autistic advantage from the accounts of adults diagnosed with autism: A qualitative study. *Autism in Adulthood, 1*(2), 124–133.
42. Kapp, S. (2013). Interactions between theoretical models and practical stakeholders: The basis for an integrative, collaborative approach to disabilities. In E. Ashkenazy & M. Latimer (Eds.), *Empowering leadership: A systems change guide for Autistic College students and those with other disabilities* (pp. 104–113). Washington: Autistic Self Advocacy Network (ASAN). Retrieved from http://autisticadvocacy.org/wp-content/uploads/2013/08/Empowering-Leadership.pdf.

Open Access This chapter is licensed under the terms of the Creative Commons Attribution 4.0 International License (http://creativecommons.org/licenses/by/4.0/), which permits use, sharing, adaptation, distribution and reproduction in any medium or format, as long as you give appropriate credit to the original author(s) and the source, provide a link to the Creative Commons license and indicate if changes were made.

The images or other third party material in this chapter are included in the chapter's Creative Commons license, unless indicated otherwise in a credit line to the material. If material is not included in the chapter's Creative Commons license and your intended use is not permitted by statutory regulation or exceeds the permitted use, you will need to obtain permission directly from the copyright holder.

14

Torture in the Name of Treatment: The Mission to Stop the Shocks in the Age of Deinstitutionalization

Shain M. Neumeier and Lydia X. Z. Brown

For Silverio Gonzalez, Abigail Gibson, Linda Cornelison, Vincent Milletich, Danny Aswad, Robert Cooper, and unknown others broken down and killed at the Judge Rotenberg Center and inside institutions everywhere—you are not forgotten and your lives were worth living.

Inhumane Beyond All Reason

Half an hour's drive south of Boston sits a facility that the United Nations has specifically condemned for its use of torture, and that Massachusetts' own state agency for people with developmental disabilities described as "inhumane beyond all reason" [1, 2]. It isn't a historical site, kept intact only as a memorial for its victims and a warning for the future. Nor is it an illegal operation that's survived through secrecy and corruption. Instead, the Judge Rotenberg Center, a self-described residential school and treatment center, continues to use an inhumane, behaviorist approach in working with youth and adults with disabilities, as it has for over forty-five years, with the open complicity of and funding from Massachusetts, New York, California, and several other states across the country [3].

The Judge Rotenberg Center's most infamous form of abuse is electric shock—a human dog shock collar in the form of a backpack that about one-fifth of its residents are forced to carry around with them throughout the day. Despite the singular attention given to this particular aspect of its program, though, the facility had been open for almost two decades before it started electrocuting the people in its care. Both its origins and its practices during that initial period make it clear that the device itself wasn't what had made the program particularly bad. Rather, if any facility were to become the only one in the country, maybe even the world, to punish autistic and other disabled people with electric shock, it would hardly be surprising for it to be one started by Matthew Israel, a protégé of infamous behaviorist B.F. Skinner who wanted to bring his mentor's fictional behaviorist utopia *Walden II* to life but needed a captive audience to do it, and one which had already killed multiple residents through abuse and neglect before [4].

The general public only became aware of and (briefly) galvanized in opposing JRC upon seeing footage—finally made public in April 2012—of a young man named Andre McCollins being repeatedly shocked while restrained face down, and this after numerous other media exposés of the practice in the decades since JRC's founding. Autistic self-advocates, and disability rights advocates more generally, have been both aware of and actively trying to shut the program down for much longer. Our opposition goes beyond the program's egregious practices in and of themselves. State agencies, state legislatures, courts, the federal government, and organizations such as Autism Speaks that claim to support us have all either refused to take a meaningful stand against JRC, its philosophy, and its practices, or have actively protected or promoted them. This speaks not only to how difficult a task it has been and will be to shut down this one alleged outlier, but to how little members of our community are valued, and therefore how significant the barriers and dangers are, or are likely to be, in other areas of advocacy.

Tearing Down the Walls They Built

Places of confinement—like residential schools, group homes, mental hospitals, and prisons—have always sought people considered weird, scary, and subversive to keep inside their walls since their advent. Our history in autistic self-advocacy and disability more generally is riddled with institutions, which we know as places of constant violence, forced treatment, involuntary medical experimentation, isolation, and layers upon layers of abuse—even and especially the ones that seem nice on the outside. There were the Fernald radiation experiments where researchers recruited institutionalized disabled children for a "science club" so they could investigate the effects of radiation by feeding them irradiated cereal without their knowledge. There were the Tuskegee experiments where researchers deliberately withheld treatment from low-income Black workers with syphilis so they could study the course of the disease. There were the developmentally disabled men confined in a bunkhouse in Iowa and forced to work for decades in a turkey slaughter factory for a subminimum wage. And there were the Willowbrooks and Pennhursts—large-scale, state-run institutions where thousands of people with disabilities were incarcerated in squalor and subject to all manner of abuses.

The history of prisons as places of confinement is old and dark, the distinction between disability institutions (places that are supposed to provide care) and penal institutions (places that are supposed to detain and punish) constantly blurring until it's hard to tell much of a difference.

Ableism is the idea that only some people's brains or bodies are healthy, whole, functional, and valuable in society, and that the rest of us are broken, defective, inferior, and unworthy. Put into practice, ableism values us based on whether we seem "normal" based on constantly-shifting goalposts, whether we work and produce according to conventional measures, and whether we can maintain the social order in a profoundly racist and classist society.

JRC and institutions like it operate knowing that most people will readily accept myths about disabled people's incompetence, inferiority, and brokenness. It's easier to dismiss us as uncontrollable, violent, and aggressive for no reason than to recognize that many of us have survived years of trauma caused by compliance training, rejection, isolation, and serial predators. It's easier to believe that nondisabled family members and "experts" know what is best for disabled people than to believe us speaking for ourselves. It's easier to lock us away instead of doing the work necessary to make sure we can all belong and exercise autonomy. At JRC, fear, revulsion, pity, and hatred pervade the place so strongly that torture can be resold as "extremely beneficial and lifesaving [treatment]" [5] that "allows [JRC residents] to integrate into the community, which is an [Americans with Disabilities Act] requirement" [6].

JRC's abuses represent some of the most extreme forms of behaviorist violence. Yet as community pioneer Mel Baggs has observed repeatedly over the past two decades, JRC is not the worst institution to have ever existed, but rather, represents thousands of institutions where staff can abuse, torture, and murder disabled people with impunity and in silence [7]. An institution may be as small as a single person, Baggs has written, so long as that person lives under the control of others [8]. Survivors of institutionalization outside JRC, including both Baggs [8] and activist and commentator Cal Montgomery [9], often describe aesthetically pleasing and seemingly progressive institutions as the most dangerous [10, 11]. In this, JRC's threat becomes clearer—disguised by flashy and bright Big Reward Store and Yellow Brick Road rooms; clean and pressed shirts and ties for residents; and newly painted group homes in the neighborhoods surrounding the main building.

Worse, JRC's marketing model holding out its ostentatiously decorated rooms as proof of its benevolence, ironically ignores that a large portion of the people confined there are much more likely to be overwhelmed and overstimulated, sometimes even to the point of physical pain, by the design of those rooms. Non-autistic people, and especially neurotypical people, also seem woefully unaware of these particular issues, despite the increase in autism "awareness" campaigns of the past several years.

JRC has also always been an atypical private institution, in its largely negatively racialized population and its constant, domineering surveillance over both residents and staff. Over the past five decades, JRC's population has shifted. Its residents were once almost entirely people with developmental disabilities with intense support needs. Now, its residents include large numbers of people whose primary neurodivergence is psychiatric disability or mental illness, many of whom arrive through referrals from the juvenile criminal legal system. According to the National Center for Education Statistics [12], in the 2015–2016 school year, JRC's school-age population was 81.5% Black or Latinx people, with all categories of people of color or racial minorities combined comprising 87.4% of its population.

This particular blend of ableist and racist targeting challenges the historically white autistic community and neurodiversity movement, by calling into question how and why so many activists working publicly against JRC have little to no understanding of the racial implications of JRC's population and increasingly overt ties to the criminal punishment system, including transfers from Rikers Island [13]. Further, those committed to anti-racism work, particularly our white and nondisabled allies, must also contend with JRC's exploitation and scapegoating of low-paid line workers who are largely immigrants of color and often the only JRC staff ever prosecuted for physical abuse (but never the shocks) while the largely white administration avoids any meaningful consequences. JRC, like all institutions, is the inevitable product of a society of prisons, which exist as a tool of social control for eradicating undesirable people and enabling appalling abuses (as punishment, treatment, or both) on those powerless to stop them from happening.

In recent years, National ADAPT, a grassroots direct action disability rights group consisting primarily of anti-institutionalization physically disabled activists, has organized multiple actions targeting JRC both in Massachusetts and in Washington, DC. ADAPT's anti-JRC work has been led in large part by the wisdom of several autistic leaders, including Anita Cameron, a proudly queer Black activist who has been organizing with ADAPT for decades, and Cal Montgomery, longtime autistic writer and activist who is also a survivor of multiple institutions. And while we haven't been able to participate in most of ADAPT's actions, we have supported and amplified their efforts in every way possible.

Now the fight against JRC is firmly in the cross-disability community's arena, and no longer the sole dominion of the small but mighty autistic self-advocacy movement, where generations of neurodiversity advocates have supported one another in fighting it. More promisingly, younger activists and advocates are now joining forces to renew the struggle against JRC and similar sites of violence, following in decades of work to tear down institution walls.

Though We Be but Small, We Are Mighty

Though we grew up on opposite sides of the country, we shared many interests and experiences that both drew us to the issue of abuse at JRC, and that made us compatible, personally and professionally. While neither of us had been institutionalized in a program like JRC nor spent significant parts of our school years in a segregated special education setting, both of us had been targeted for disability-related harassment and discrimination by school officials as well as peers. As bold, outspoken, unconventional people from the start, we had frustrated and been frustrated by neurotypicals who'd valued compliance and conformity for their own sake and seemingly above all else, and whatever victories we might have won in these conflicts tended to be moral rather than actual.

Both of us also had tendencies to become interested in, and by neurotypical standards obsessed with, dark and violent subject matter, but from a perspective of wanting to solve the problems that we saw. Lydia, for instance, had had a longstanding interest in the terrorist attacks of September 11, 2001, and the U.S. government's subsequent repression of Arabs and Muslims as part of its so-called War on Terror. And while by this point Shain was no longer as fixated on the horrors of animal experimentation as they had been as a child, they'd continued exploring themes of abuse and oppression through fiction for lack of concrete ideas on how to confront them in reality.

We joined the decades-long fight to end aversives and close JRC in 2009, around the same time that we were both entering the autistic activist community in Boston. Lydia, who had grown up in the Boston metro area and known they were autistic since early adolescence, was a member of the

Autistic Self Advocacy Network (ASAN) Boston chapter. By then a high school student, they were already drafting, then introducing, a bill in the Massachusetts state legislature that would have mandated police training on autism. Lydia had connected with ASAN when they learned about the neurodiversity movement from autistic activists' criticisms of the charity Autism Speaks, and began reading blogs by activists like Bev Harp (Square 8, aspergersquare8.blogspot.com), Mel Baggs (Ballastexistenz, ballastexistenz.wordpress.com), and Kassiane Asasumasu (Radical Neurodivergence Speaking, timetolisten.blogspot.com). Lydia's early activism included organizing mass opposition to criminalization and restraints targeting autistic students in Arizona, Alabama, and Kentucky.

Shain, meanwhile, had spent much of their childhood undergoing, recovering from, or trying to avoid involuntary psychiatric and surgical treatment. However, they only learned they were autistic as an adult, at about the same time that they were applying to law school. Although they had initially joined online message boards like WrongPlanet and Aspies for Freedom to learn more about their autistic identity and find friends, they soon ran across information on the widespread, systemic abuse that autistic and other disabled people face in the name of treatment.

It was in this context, and in the aftermath of several recent and highly publicized exposés of abuse at JRC, that we separately became aware of the abuse going on in our own figurative backyard. As with other things in both of our lives that had horrified and fascinated us, we started painstakingly collecting information on the issue, then acting in whatever way became apparent—writing articles, speaking publicly, testifying at hearings—first on our own, and then, increasingly, collaboratively.

New Resistance and Organizing Against the School of Shock

While attending law school in Boston, Shain threw themselves into challenging JRC, confronting Massachusetts Governor Deval Patrick about JRC during a public appearance, authoring two papers on legal strategies to stop JRC, leading a session at the annual Rebellious Lawyering Conference, and giving an invited presentation at the Symposium on Ethical, Legal, and Social Implications of Autism Research.

In April 2012, Shain attended nearly the entire medical malpractice trial against JRC for torturing Andre McCollins, sitting only feet behind Matthew Israel while taking painstaking notes for the public [14–20]. During that trial, video of JRC's shocks first aired publicly.

Shain also began working with troubled teen industry survivors, including with the Community Alliance for the Ethical Treatment of Youth (CAFETY). While at CAFETY, Shain took a leading role in organizing what would be one of the largest anti-JRC protests in the next decade. In July 2012, outraged by the McCollins video, hundreds of activists, many forced treatment and institutionalization survivors, gathered in Boston for a State House rally, then reconvened outside JRC, marching through rain and barricades.

In Fall 2012, Shain moved to Washington, DC to work as CAFETY's policy associate. Shain took the lead role in drafting and presenting a report on institutional abuse targeting youth for the United Nations Special Rapporteur on Torture's expert consultation on torture in healthcare settings. There, Shain connected with Lydia, who authored a follow-up submission on JRC on ASAN's behalf. Lydia's report [21] later turned into an article, "Compliance is Unreasonable: The Human Rights Implications of Compliance-Based Behavioral Interventions under the Convention Against Torture and the Convention on the Rights of Persons with Disabilities," published in a compilation edited by the U.N. Special Rapporteur on Torture [22]. ASAN meanwhile invited Shain to author a brief history of JRC for its groundbreaking anthology on the neurodiversity movement [1].

During the same period, Lydia worked alongside advocates from ASAN Boston to lobby Massachusetts lawmakers for measures to limit and stop JRC's abuses. Lydia also began to curate information about JRC for a dedicated page on their blog Autistic Hoya, which would later become the JRC Living Archive and Document Repository (https://autistichoya.net/judge-rotenberg-center).

Throughout 2012 and 2013, former JRC employee Gregory Miller wrote a series of widely publicized anti-JRC essays [23, 24]. He described vicarious trauma from witnessing abuse and being coerced to participate before realizing the full extent of the harm he was responsible for, his letters to JRC condemning it, and his resignation. Miller's Change.org petition [23] amassed over 200,000 signatures, and incited another rally at the Massachusetts State House. There, Miller spoke alongside Cheryl McCollins, who by now had been barraged with constant exposure to the video of her son's torture. The year saw three anti-JRC rallies before the July 2012 demonstration. Meanwhile, Lydia worked with Miller and others to present about JRC's abuses for various autism and disability advocacy organizations.

In January 2013, we planned a demonstration against the U.S. Food and Drug Administration's reticence to regulate the shock devices, set at its Maryland headquarters. The night before, we huddled in a college dorm building while creating colorful signs—Stop the Shocks, People Not Experiments, No Compromise on Torture, Disability Rights are Human Rights, Ban the GED (JRC's shock device). Three others joined us on the traffic island across from the entrance. It was a lonely day for us five, and nearly as many Homeland Security police cars arrived to watch us.

One month later, in February 2013, Lydia received an email containing a message from a survivor hoping to discuss JRC. Lydia replied to the letter writer, received permission to publish it anonymously, and began circulating it to increase consciousness of JRC's abuses [25]. Roughly simultaneously, Massachusetts quietly filed a motion to void the 1987 court order enshrining legality of the shocks.

Come 2014, and rumors that FDA officials would finally consider banning electric shock aversives, we were split with Shain in Oregon and Lydia in Jordan. The FDA announced a public hearing on the possibility of banning the shocks in April 2014. Shain mounted a campaign to raise funds to cover their and later Lydia's travel to present testimony. Before our flights, we worked for two nights across continents on our first anniversary preparing detailed, heavily-cited comments in attempts to pre-emptively bury JRC's supporters in research.

Amid hours of testimony from advocates who'd worked in a coalition bridging policy, research, and activism, and JRC's sometimes screaming supporters, survivors Jennifer Msumba and Ian Cook commanded full attention from all present. Msumba's testimony, delivered by video, described intense pain, burns, and post-traumatic stress disorder caused by repeated shocks to punish and control her. Cook opened his testimony by announcing that since leaving JRC, he has come out as transgender, noting defiantly JRC's use of his deadname while confined there. His conclusion could compel no response from JRC's supporters—"I was in an abusive relationship two years ago, and part of why I fell prey to it is that JRC instilled a lesson in me that it is okay for people to hurt me so long as they are trying to correct me" [26].

Since that hearing, we have collaborated to support many other disabled people in organizing direct action and submitting testimony against the torture, drawing constantly on the leadership of survivors like Msumba, Cook, and Terri Du Bois, who have all spoken out against the horrors they survived and witnessed.

In October 2014, sensing declining interest in anti-JRC activism, Lydia hosted Shain and Msumba for a panel on institutional abuse targeting disabled people, which received modest press coverage amid announcements of Msumba filing a lawsuit. In August 2015, we organized community testimony against JRC for the perennial Massachusetts hearing, but once again, the legislature refused to act. In December 2015, we presented on JRC and other institutions' abuses to a packed room at a national conference.

In April 2016, two years after its hearing, the FDA finally announced a proposal to ban the shocks. Along with many others, we pushed for massive public pressure to finalize the regulation as law, but the FDA never made a decision. By 2017, the presidential administration changed amid conflicting indications about federal agencies' desires to either accelerate or severely delay their regulation-making powers. Apart from ADAPT's large-scale public actions in October 2016 (at JRC) and March 2018 (targeting the FDA in Washington, DC, and Maryland), public attention to and interest in JRC has largely faded.

In June 2018, when a Massachusetts probate court judge ruled in JRC's favor in the lawsuit stemming from 1987, one major door closed.

Freeing (All) Our People

Even after what seemed like an increase in the awareness of and activity to end abuse at JRC, it remains open. Worse, restraint, seclusion, food deprivation, physical assaults, abusive behavior modification therapies, and institutionalization of disabled people remain legal and widespread outside of JRC.

There is some reason for hope of progress at this point, at least as it concerns JRC in particular. As of the time this piece was written in early 2019, the U.S. Food and Drug Administration recently announced that it plans to finalize the proposed ban on the use of shock devices. Furthermore, although JRC won the most recent legal battle against the state of Massachusetts over the government's attempts to ban or even just limit the use of aversives, the state is in the process of appealing this decision. JRC has also lost its staunchest ally in the state legislature after former representative Jeffrey Sanchez, whose nephew Brandon has been at the facility for decades, was defeated in the 2018 primary elections. Meanwhile, there are both ongoing and forthcoming lawsuits by survivors and their families seeking justice, as well as by disability advocates hoping to bring about systemic change.

At the same time, though, there have been several recent instances of JRC staff being caught hitting or beating the people in their care that have resulted in criminal investigations and convictions. These incidents demonstrate that JRC's culture of abuse goes much deeper than its use of electric shock, and that even assuming the shock ban comes into effect, advocates will need to continue to press for JRC, along with other, similar breeding grounds for abuse, be investigated, defunded, and ultimately shut down.

While policy advocacy and lawsuits are two avenues through which to work toward these goals, the effectiveness of laws still hinges on the framework in which they're written, the ways in which they'll be interpreted, and the stringency with which they'll be enforced by judges, licensing agencies, and other decision-makers. Were it possible to get a law or policy banning the use of not only electric shock devices but all the forms of aversives JRC has used as part of its behavior modification program, this would still not fully address the core problems that JRC represents. Namely, even many opponents of the egregious types of aversives used there still see behavior modification aimed at making autistic people be more compliant as worthwhile. Nor is there nearly as widespread a rejection of institutionalization as a whole except within small and still relatively powerless communities of dedicated advocates. To create a society in which not only is there no JRC, but also nothing remotely comparable, these more accepted goals and practices have to be challenged just as unequivocally as shock devices.

The good news is that this isn't just a job for lawyers, lawmakers, and protesters, or even for others with skills like writing letters or making phone calls that are often associated with political advocacy. While people in these roles can and should continue to lend our skills whenever possible, it also falls to educators and service providers to challenge the beliefs and systems surrounding autistic youth that allow abuse up to and including the type that occurs at JRC to continue. These same professionals can also change their own practices to honor the autonomy, dignity, and humanity of the people they work with, and train their colleagues to do the same. Likewise, parents of autistic people have the opportunity and responsibility to use their voices as culturally recognized authorities on autism to defend their children against coercive, abusive attempts to make them comply with neurotypical norms for their own sake. Researchers can also shape the types of interventions that are further explored, funded, or abandoned by studying not only what interventions are most "effective" but what they're most effective at, and conversely, what outcomes are worth effecting for the well-being of the people most directly involved. Meanwhile, journalists and artists can shape cultural narratives around disability, shifting them away from their current direction of encouraging a return to confinement and forced treatment in institutions and instead toward one that will make even more common forms of abuse seem unimaginably horrific within a generation.

The results of these efforts would go beyond the absence of coercion and abuse, though. In concrete terms, these positive changes could, should, and must include a service delivery system that's truly directed by neurodivergent people in every sense. For instance, to the extent that any sort of congregate care or living facilities would still exist, they would be peer-run, non-hierarchical, and truly voluntary. Service recipients would have a meaningful ability to leave, choose different supports, or refuse placement in one at any point, without caretakers being able to override this decision or agencies being able to deny them services in their homes and communities. This would require directing resources away from institutional facilities and coercive practices, and creating a new system that prioritizes and in fact guarantees community integration and self-determination.

While most of these approaches and solutions can't shut down JRC on their own, they will be crucial in creating a society in which it's impossible for any place like it to exist, and more generally, where autistic people can live safely and on our own terms.

References

1. Neumeier, S. (2012). Inhumane beyond all reason: The torture of autistics and other disabled people at the Judge Rotenberg Center. In J. Bascom (Ed.), *Loud hands: Autistic people, speaking* (pp. 204–219). Washington, DC: The Autistic Press.
2. Cabral, D. (1995). *Investigation report* (p. 25). Massachusetts Department of Mental Retardation. Retrieved from https://autistichoya.files.wordpress.com/2016/04/mass_dmr_lc_investigation.pdf.
3. Massachusetts Department of Developmental Services. (2017, August 17). *Re: Public Record Request—July 26, 2017* (Letter to Shain Neumeier).
4. Massachusetts Disabled Persons Protection Commission. (1995, January 3). Investigation Report (Agency Case No. 2540-94-076, DPPC Case No. 12440).
5. Shear, M., & Shear, M. (2014, June 24). The FDA may ban the treatment keeping our daughter alive. *The Washington Post*. Retrieved from https://www.washingtonpost.com.
6. Open Public Hearing of Food and Drug Administration, Medical Devices Advisory Committee, Neurological Devices Panel, 98–107 (2014a). *Testimony of Nathan Blenkush*. Retrieved from https://autistichoya.files.wordpress.com/2016/04/fda-neuro04-24-14-final.pdf.
7. Baggs, A. M. (2006, December 6). *Why students praise the Judge Rotenberg Center* (Web log post). Retrieved from https://ballastexistenz.wordpress.com/2006/12/06/why-students-praise-the-judge-rotenberg-center.

8. Baggs, A. M. (2012, January 23). *What makes institutions bad* (Web log post). Retrieved from https://ballastexistenz.wordpress.com/2012/01/23/what-makes-institutions-bad.
9. Montgomery, C. (2001). Critic of the Dawn. *Ragged Edge Online.* Retrieved from http://www.raggededgemagazine.com/0501/0501cov.htm.
10. ADAPT. (2018, March 15). *Pain and fear teach nothing: An ADAPTer reflects on the Judge Rotenberg Center* (Press release). Retrieved from https://adapt.org/press-release-pain-and-fear-teach-nothing-an-adapter-reflects-on-the-judge-rotenberg-center.
11. Baggs, A. M. (2014, June 09). *This is how I feel when I read a lot of posts about the Judge Rotenberg Center* (Web log post). Retrieved from https://ballastexistenz.wordpress.com/2014/06/09/this-is-how-i-feel-when-i-read-a-lot-of-posts-about-the-judge-rotenberg-center.
12. National Center for Educational Statistics. (2016). *The Judge Rotenberg Educational Center* (ID A9701991). PSS Private School Universe Survey Data for the 2015-2016 School Year. Retrieved from https://nces.ed.gov/surveys/pss/privateschoolsearch/school_detail.asp?Search=1&ID=A9701991.
13. Gonnerman, J. (2007, August 20). Nagging? Zap. Swearing? Zap. New York's investigations of the Rotenberg Center. *Mother Jones.* Retrieved from https://www.motherjones.com.
14. Neumeier, S. (2012, April 16). *The Judge Rotenberg Center on trial: Part one* (Web log post). Retrieved from https://autisticadvocacy.org/2012/04/the-judge-rotenberg-center-on-trial-part-one.
15. Neumeier, S. (2012, April 17). *The Judge Rotenberg Center on trial: Part two* (Web log post). Retrieved from https://autisticadvocacy.org/2012/04/the-judge-rotenberg-center-on-trial-part-two.
16. Neumeier, S. (2012, April 18). *The Judge Rotenberg Center on trial: Part three* (Web log post). Retrieved from https://autisticadvocacy.org/2012/04/the-judge-rotenberg-center-on-trial-part-3.
17. Neumeier, S. (2012, April 25). *The Judge Rotenberg Center on trial: Part four* (Web log post). Retrieved from https://autisticadvocacy.org/2012/04/the-judge-rotenberg-center-on-trial-part-4.
18. Neumeier, S. (2012, April 25). *The Judge Rotenberg Center on trial: Part five* (Web log post). Retrieved from https://autisticadvocacy.org/2012/04/the-judge-rotenberg-center-on-trial-part-5.
19. Neumeier, S. (2012, April 26). *The Judge Rotenberg Center on trial: Part six* (Web log post). Retrieved from https://autisticadvocacy.org/2012/04/the-judge-rotenberg-center-on-trial-part-6.

20. Neumeier, S. (2012). *The Judge Rotenberg Center on trial: Part seven* (Web log post). Retrieved from https://autisticadvocacy.org/2012/04/the-jrc-on-trial-part-7.
21. Brown, L. X. Z. (2012). *Compliance-based behavioral interventions for disabled People as cruel, inhuman, and degrading treatment and torture* (Report to United Nations Special Rapporteur on Torture). Washington, DC: Autistic Self Advocacy Network.
22. Brown, L. X. Z. (2014). Compliance is unreasonable: The human rights implications of compliance-based behavioral interventions under the Convention against Torture and the Convention on the Rights of Persons with Disabilities. In J. E. Méndez & H. Harris (Eds.), *Torture in healthcare settings: Reflections on the special rapporteur on Torture's 2013 Thematic Report* (pp. 181–194). Washington, DC: Anti-Torture Initiative, The Center for Human Rights & Humanitarian Law, American University Washington College of Law.
23. Miller, G. (2012). *Judge Rotenberg Educational Center: Please stop painful electric shocks on your students* (Petition). Change.org. Retrieved from https://www.change.org/shock.
24. Miller, G. (2013, January 16). *Letter from former teacher at Torture Center*. Autistic Hoya. Retrieved from https://www.autistichoya.com/2013/01/letter-from-former-teacher-at-torture.html.
25. xxx. (2013, January 15). *Judge Rotenberg Center survivor's letter* (Web log post). Retrieved from https://www.autistichoya.com/2013/01/judge-rotenberg-center-survivors-letter.html.
26. Open Public Hearing of Food and Drug Administration, Medical Devices Advisory Committee, Neurological Devices Panel, 207–208 (2014b). *Testimony of Ian Cook*. Retrieved from https://autistichoya.files.wordpress.com/2016/04/fda-neuro04-24-14-final.pdf.

Open Access This chapter is licensed under the terms of the Creative Commons Attribution 4.0 International License (http://creativecommons.org/licenses/by/4.0/), which permits use, sharing, adaptation, distribution and reproduction in any medium or format, as long as you give appropriate credit to the original author(s) and the source, provide a link to the Creative Commons license and indicate if changes were made.

The images or other third party material in this chapter are included in the chapter's Creative Commons license, unless indicated otherwise in a credit line to the material. If material is not included in the chapter's Creative Commons license and your intended use is not permitted by statutory regulation or exceeds the permitted use, you will need to obtain permission directly from the copyright holder.

15

Autonomy, the Critical Journal of Interdisciplinary Autism Studies

Larry Arnold

The dominant discourse in Autism since the first appearance of the word in the psychiatric literature, has been what has subsequently been called the Medical Model [1] but in recent times there have been many challenges drawing from the field of disability studies and the emerging field of critical autism studies. This is the story of how I came to start *Autonomy, the Critical Journal of Interdisciplinary Autism Studies* [2].

I did not discover my autism—I prefer the word discovery over diagnosis—during the lifetime of either of my parents. I had not felt any great need for an identity whilst they were still alive and put off many questions I perhaps ought to have addressed whilst they were still there to answer them.

I had been long aware that I related to the world in a very particular way, when I watched others around me as I grew up, negotiate the world with an apparent ease that was foreign to me. In my mid-twenties I took

L. Arnold (✉)
Coventry, UK
e-mail: lba657@alumni.bham.ac.uk

on the care of my mother, who was becoming increasingly dependent upon on my help. Eventually she would become completely dependent on a wheelchair for mobility. This roughly coincided with an increasing awareness of disability rights following the International Year of Disabled People in 1981, which first became a focus for our activities. We both had a passion for social justice, she being active in Women's rights and myself being a member of the Labour Party since I left University. She would make public speeches and sit on committees, whilst in addition to transporting her to meetings I would help her to prepare for them through research.

When she passed on, it was not as if the focus for that had gone, but that for the first time I was left on my own to negotiate the social world on which so much of her activity was predicated.

Diagnosis eventually came at the point of crisis, as it often does to both adults and teens. For me the crisis was a middle-aged transition with both parents gone when I first became aware of my fragile status as an adult with no family to base my identity around. It was for me every bit as traumatic as the transition from school to work is for autistic youth.

The foundations of my current advocacy had been long in the making though, something that my mother had encouraged me in when I was struggling with unemployment more than a decade and a half earlier. She thought rightly that I needed a focus in life, some kind of structure to prevent my life from falling apart, and she encouraged me to join her on a course in disability rights.

That is where my purpose started, and although she was unaware that I would later attract the label of autism she had sufficient knowledge of what I needed at that time to give me the confidence to deal with the public at large.

As a disabled person, she too had become isolated from the "mundane" world, shunned and rejected by people who were embarrassed and ignorant of how to relate to disabled people. Our social circle increasingly revolved around newly found disabled friends. These were people I felt most comfortable to be among, and who showed the most understanding of difference.

The 1980s were a time when the social model of disability was developing, and my mother had challenged me to accept myself as disabled. She realized through her experiences with other parents, that I would by later definitions have been considered as having "special needs" when growing up. The model gave us a powerful tool to confront the inequalities and disabling mores of society by seeing the problem as not being inherently within, but caused by the political and economic systems to accept and adapt for difference. All of the seeds which led me to start *Autonomy* were there at the time. Not just the ideological tools, but some practical ones too. I was an early adopter of computer technology, which I found liberating, and I used it to the full. I used to compile databases of information. This led to my compiling information and publishing it in several editions of a directory of services provided by the City Council, National Health Service, and Department of Social Security.

Whilst the directories told people of what was available, I used those same skills to campaign for what was not, and I started a newsletter for the Coventry Council of Disabled People, an organization my mother and I helped to found in 1983. Thus I learned the skills of editing and word craft that I later put to such use as I have in academia and elsewhere.

Neurodiversity

Neurodiversity itself was not a concept I discovered until after my mother had passed on, however it was something I understood from the medium of disability studies nonetheless, in that I first read Judy Singer's [3] article in its context as a contribution to a compilation of emerging critiques of existing disability models.

As an early adopter as it were, using computers since the mid-1980s, I finally took the plunge into the Internet in 1996 where I started my exploration into the world of 'neurodivergent' identity. I found others like myself on various web sites, mailing lists, and newsgroups. I expect without them I would have remained isolated and unaware but by 1997 I had a website of my own (http://www.larry-arnold.net/), and my first domain not long after.

My first practical steps in the world of neurodiversity outside of the Internet came when I started to organize a local meet up for dyspraxic people under the auspices of the late Mary Colley and the adult group of the Dyspraxia Foundation which I had been encouraged to join.

It was a world of autistics and cousins, a terminology I discovered on joining Jim Sinclair's ANI-L (the mailing list for supporters of Autistic Network International, the organization that Jim had founded with Donna Williams and Kathy Xenia Grant). There were many of us who had multiple labels of dyspraxia, Tourette's syndrome, dyslexia, epilepsy, Asperger's syndrome and autism so "Neurodiversity" seemed to be a convenient banner to unite under, and I founded the Coventry and Warwickshire Neurodiversity Group, what may well be one of the first organizations to rally under the name of neurodiversity. We were a breakaway from a group of students run along the lines of a support group but by a psychologist with a failure to understand the need for personal "autonomy."

I was finding for myself a new role where I could continue the advocacy I had begun with my mother, in support of a community I increasingly felt a sense of being at home among. I took it up with a passion and zeal and Jim Sinclair's writings had a profound effect. They were, as I described them recently, foundational documents, our Declaration of Independence as it were. I did not want to see them lost to posterity because of the ephemeral nature of the World Wide Web.

In 2003 Mary Colley formed a national group under the Neurodiversity heading called the Developmental Adult Neuro-Diversity Association or DANDA for short [4]. This was another important first for neurodivergent-led and—controlled organizations. Although I had differences with Mary over the redefinition of Neurodiversity as purely "developmental" I was one of several people involved with DANDA who went on later to challenge the National Autistic Society (NAS) from the perspective of the well-used disability rights motto "nothing about us without us."

My claim to fame was in breaking the glass ceiling of that society in becoming the first diagnosed autistic person to serve on the board in 2003. Not I humbly add, the first autistic person to serve on the board of an autism charity—both Thomas McKean and Stephen Shore served on boards in the USA—but the first to make a major impact on the direction of the largest autism charity in the UK. For all that autistic people still have their concerns about the NAS, I believe it is vastly different from what it might have been had I not made my presence felt.

It was around the time that I had become involved in the NAS that I started to go to conferences. I will call them "conferences about autism" rather than "autistic conferences" because the autistic input if it was there at all, was minimal and confined to what Jim Sinclair has called the "self-narrating zoo exhibit" phenomenon, where the only role open is tokenistic, and the only justification in the organizer's eyes is to talk to the non-autistic audience about how awful it is to be autistic.

I also began to hear the so-called experts on autism speak, and to ask myself "Are they talking about us?" because it did not sound like they were describing the people I had come to know increasingly as autistic in our world. I would suppose a key moment was when I heard somebody ask autism laureate Uta Frith, if she knew whether the sensory sensitivities observed among autistic children persisted through adulthood. She answered that she did not know, "Perhaps they grow out of them" she said. At which point, I, a strapping autistic "youth" of some 46 summers could contain myself no longer. "Not for me they didn't" I called out, not the last interjection of mine to that conference either.

At this point I need to take a couple of steps back to look at those other parts of the roadway that were leading me toward the establishment of *Autonomy*.

Academia

In 2002 I discovered that the University of Birmingham was pioneering an Internet-based degree course for professionals involved with, and parents of, autistic children. I thought "Why should it not be open for autistic people too?" In the same spirit in which I set about to challenge the NAS from within, I set about to change the course from within the University. I became the first of one of a select few autistic people, along with Claire Sainsbury, David Andrews, and Heta Pukki, to graduate from Birmingham and in so doing we opened the doors for many others to follow. It wasn't easy being among the first; it never has been.

That being said, it was with growing confidence when I moved from taught master's level studies to Ph.D. research of my own. I had found my niche, and more doors opened for me to engage with academics far and wide on a level playing field as a bona fide researcher, not just a conference attendee with a list of awkward questions.

I had in addition to my autism qualifications, a vocational Higher National Diploma in Media Studies, Moving Image, which I had been studying at the same time as the Web autism course. This was quite a feat as I was studying Psychology at a third college in the evenings as well. Toward the end of media course, I produced a commercially available video with one of the staff there who was also studying media himself. I followed it up afterward with a second DVD, with the same collaborator. The first video was all too much "self-narrating exhibit", but the second one was based on a presentation I had made at the first ever Autscape conference in 2005. This time the video was addressing important questions about the representation of autism, and questioning the diagnostic categories of autism and Asperger's syndrome as they then existed in DSM-IV-TR psychiatric manual [5]. It was perhaps another attempt to talk back to the non-autistic people who were defining us and it was very appropriate material for the first conference/retreat organised in the UK by autistic people, for autistic people. Autscape took its cue from Autism Network International's Autreat conferences which Jim Sinclair had organized in the USA. I have since seen the video described in an academic thesis as "an important autoethnography in this field" [6].

So everything was beginning to come together. I had experiences as a publisher and as an editor since the mid-1980s (service directories and newsletter), and also as an academic presenter and lecturer since 2005. I had become in every sense an academic and engaged beyond pure advocacy into the realms of academic matters which defined the very way in which professionals and clinicians talked about us. I still had a deep and prevailing sense of dissatisfaction with the whole manner in which academia referred to us and continued to consider it as a form of exclusion in which it was still largely a discourse about us without us.

I had answered back with my video, and with my conference presentations but it was increasingly clear that the two worlds which CP Snow had described in the 1950s where humanities did not understand science, and science had no grasp of the humanities was still the paradigm for today. On the one hand, I was familiar and had engaged with the medical/psychiatric researchers, and on the other with the worlds of sociology and disability studies and I began to wonder whether we were really getting that much respect from either. It was also a question of ethics, not conventional ethics as in getting your proposal past the ethics committee but more teleological in the sense of whom did the research serve.

So summing up, the factors leading me toward "Autonomy" were: a foundation in disability rightsand mental health advocacy from a social modelperspective; experience of editing and publishing; and experience of academia as a student, as a researcher, and as a presenter. What else did I need other than the will to start it?

The Journal

Eventually, whilst still completing my Ph.D., I thought I might as well go for it and start a journal of my own. I had managed already to capture the essence of video production and publication, and I knew I had the capacity to learn whatever I needed to realize what I wanted to do. I wanted it to be something more than merely another version of the many websites and blogs already on the Internet; I wanted something that looked and behaved like the established journals which academics are used to reading and contributing to. I looked around and found that open publishing was the way to go, and determined that the Open Journal Systems platform which had been developed for this purpose by the "Public Knowledge Project" [7] was the best one for me to use, being as it was free to download and had the right tools to create a professional-looking journal that could sit easily among the existing online journal formats.

I determined from the start that it should largely be a peer-reviewed journal, in order to give it the same academic status as journals such as the *Journal of Autism and Developmental Disorders*, and *Good Autism Practice*. However, I also allowed scope to include articles that had never been published or written in an academic context but which could still be considered, as I said before, as "foundation documents" of our autistic movement and community.

The choice of name, *Autonomy*, suggested itself because not only does it encapsulate what the thrust of my thinking had been, agency and autonomy for autistic people, it contains the same "Aut" root of Autism, which has been used before for organizations and events such as Autreat, Autscape, and Autreach. Of its subtitle (*the Critical Journal of Interdisciplinary Autism Studies*), "Critical" was to embody a questioning and examining of the prevailing paradigms of autism research. "Interdisciplinary" indicated that it wished to include contributions from a variety of academic fields.

I endeavored from the beginning to get some support from established and respected Autistic academics such as Stephen Shore and Temple Grandin, who both agreed to lend their approval to the journal. Closer to home, I sought help and general advice in my editorial decisions from Dinah Murray and Damian Milton, both of whom were associated with the University of Birmingham at thetime.

At first I was wary of pushing the Birmingham connection. I had incorporated the University as part of the masthead design for the journal and when the head of the college of social sciences asked to have a private word with me, I was worried that I had committed another academic faux pas. Fortunately it turned out that he wished me to introduce the journal at a plenary session of the forthcoming education graduate conference. I was even presented with an award for my contribution to the research community on the basis of my efforts to set it up.

The Ethos of "Autonomy"

In the words of the rubric: "The emphasis will be on encouraging contributions from autistic scholars who have hitherto had limited exposure to academic publishing. We will feature papers reviewed by respected academics in the appropriate fields, reviews and also feature an opinions section which it is hoped will stimulate a lively interdisciplinary debate. 'Autonomy' will appeal to the widest range of the current autism research community and foster cross disciplinary discourse between the fields of medical research, education and sociology amongst others."

This means in practical terms it encourages, but does not limit itself to autistic contributors alone, as it is important to foster debate and be inclusive. Contributors do not have to declare that they are autistic as there are a number of academics, who for a variety of reasons would find full disclosure difficult in their working environments.

However one early aim of the journal was to highlight and republish pieces that were not written from an academic perspective, but an autistic one. These articles were perhaps familiar to the autistic community but in danger of becoming lost, forgotten or difficult to find as the Internet grows and changes. They were written by people who have had important and pertinent things to say about the autistic community and who have certainly been influential in the development of autistic advocacy, Jim Sinclair's "Why I Dislike 'Person First' Language" [8] and "Don't Mourn for Us" [9] being two examples. I sought to bring these within and alongside the academic canon on the basis of merit, by giving them publication in an academic journal.

"Autonomy" is not without controversy however. Other academics since the debut of *Autonomy* have staked claims to Critical Autism Studies, and some of these have been antagonistic to the original claim, in that they are predominantly non-autistic-led discourses. In the interests of dialogue, it follows that not every article in *Autonomy* will have been written by an openly autistic author. There is room for allies, and for those who are uncomfortable with sharing their status in the public domain for fear of professional repercussions. The main focus however, is on respect toward the autistic community of scholars and the intellectual ownership of ideas that originated within Autistic communities.

References

1. Oliver, M. (1990). *The individual and social models of disability.* Paper presented at Joint Workshop of the Living Options Group and the Research Unit of the Royal College of Physicians. On People with established locomotor disabilities in hospitals July 1990. Retrieved May 10, 2010, from http://www.leeds.ac.uk/disability-studies/archiveuk/Oliver/in%20soc%20dis.pdf.

2. Arnold, L. (2012). Editorial. *Autonomy, The Critical Journal of Interdisciplinary Autism Studies*, 1(1).
3. Singer, J. (1999). Why can't you be normal for once in your life? In M. Corker, & S. French (Eds.), *Disability discourse*. Buckingham: Open University Press.
4. Arnold, L., Milton, D., Beardon, L., & Chown, N. (2018). England and autism. In F. R. Volkmar (Ed.), *Encyclopedia of autism spectrum disorders*. New York: Springer.
5. APA. (2000). *Diagnostic and statistical manual of mental disorders, edition four* (various ed.). Washington, DC: American Psychiatric Association.
6. Ellis, S. J. (2014). *Perspectives of the autistic 'Voice': An ethnography examining informal education learning experiences*. Sheffield: Sheffield Hallam University Research Archive.
7. Public Knowledge Project. (2014). *Open Journal Systems*. Retrieved March 25, 2018, from Open Journal Systems https://pkp.sfu.ca/ojs/.
8. Sinclair, J. (1999). *Why I dislike person first language*. Retrieved May 20, 2010, from http://web.archive.org/web/20070715055110/web.syr.edu/~jisincla/person_first.htm.
9. Sinclair, J. (1993). Don't mourn for us. *Our voice, the newsletter of Autism Network International*, 1(3).

Open Access This chapter is licensed under the terms of the Creative Commons Attribution 4.0 International License (http://creativecommons.org/licenses/by/4.0/), which permits use, sharing, adaptation, distribution and reproduction in any medium or format, as long as you give appropriate credit to the original author(s) and the source, provide a link to the Creative Commons license and indicate if changes were made.

The images or other third party material in this chapter are included in the chapter's Creative Commons license, unless indicated otherwise in a credit line to the material. If material is not included in the chapter's Creative Commons license and your intended use is not permitted by statutory regulation or exceeds the permitted use, you will need to obtain permission directly from the copyright holder.

16

My Time with Autism Speaks

John Elder Robison

I didn't know much about autism when I began my journey as an advocate. What I knew was my own life, much of which had felt pretty crummy. Yet I had stayed the course, leaving home, learning to make a living, and figuring out how to be an adult in America. At some point I realized I had done ok, despite my marginal childhood, and I started looking for a way to give something back to the community.

At the time my sense of community was local; defined as the area in Western Massachusetts where I'd grown up. By the early 90s I was in my 30s, with a wife and a young son, and I knew there must be millions of young people growing up marginalized, as I had been. Some were abused, others abandoned. I wondered who spoke to them, and if anyone told them life can get better when we grow older. That was the start of my advocacy.

A friend of a friend invited me to a school where I talked with at-risk kids. Another friend invited me to the local jail where I met people in a

---

J. E. Robison (✉)
Amherst, MA, USA

pre-release program. I made several unforgettable visits to Brightside, a Catholic organization that sheltered teens who were victims of abuse and neglect. Having grown up in those circumstances myself I understood the pain of those kids.

My message was that we can move beyond childhood traumas and become successful adults. We are not predestined for jail or the street. Even without college credentials (which were out of reach for me and many of them) I'd been successful (at building a business repairing and restoring cars), and if I could find a wife, get a job, or establish a business, they could too.

In the midst of that advocacy I learned I am autistic. A therapist who'd gotten to know me shared that insight, essentially out of the blue. That was a stunning discovery for me. For the first time, I was presented with a non-judgmental explanation for so many of the challenges of my life. Later I would come to see how autism didn't just disable me—it also helped me with unusual powers of focus, concentration, and sensory sensitivity.

When I first heard I was autistic, I was disbelieving because I imagined autism as total disability and I didn't see myself that way. But as I read Tony Attwood's book *Asperger Syndrome* [1] I realized the therapist was right. The description from the book was me, point by point, and that insight was enough to open my eyes, and begin a process of self-improvement that continues today. That informal diagnosis has been confirmed by the Autism Diagnostic Observation Scale [2] and other processes at autism clinics. All opened windows into my mind, and how I'm the same and different from others around me.

Today I know that people of my generation were seldom diagnosed with autism if we could talk, and there is a whole generation of people like me, who grew up without a proper diagnosis. In school we were said to be emotionally disturbed, oppositional, lazy, or stupid. The problem was, those descriptors didn't lead to therapies that were very useful for someone like me. They also predisposed others to a rather negative view of individuals who were "different."

Once I learned about autism I realized there must be many other young autistics just like me. I felt I had a message for them, but did not know how to find them. In 2006 I decided to share my thoughts in a book. That narrative became *Look Me in the Eye* [3] and its publication connected me to autistic people all over the world.

Readers looked to me as an expert on autism, but I wasn't an expert in the traditional sense. I had never studied autism in the way a teacher or psychologist might. Yet I had a lifetime's experience being autistic. To the extent my traits were characteristic of autistic people, I had an inside understanding of them.

Today autistic people are visible everywhere, but that was not the case a decade ago. We existed in the same numbers but we were invisible. Most adults were like me—undiagnosed. Children and adults who were diagnosed were often ashamed because autism had the reputation of being a terrible disability. Few were willing to step forward and say, Look at me! I'm autistic!

Yet some people did just that. Daniel Tammet released a bestselling book (*Born on a Blue Day*) [4] about being autistic a few months before my own came out. Temple Grandin and Margaret Scariano [5], Donna Williams [6], and Stephen Shore [7] had published stories previously. All of us were unique in terms of our interests and abilities, but we had this in common: We recognized that autism was a way of being, not a disease to be cured, and we should make our best life as autistic adults.

That viewpoint stood at odds with an emerging community of parents whose kids were being diagnosed in increasing numbers. Changes in the diagnostic standards and evolving awareness resulted in an explosion of diagnoses, and at the time, many assumed autism itself was becoming an epidemic. Some parents seized on the idea that their kids were injured by vaccine, and they talked about cure and prevention.

When *Look Me in the Eye* went on sale it competed with another newly released book, *Louder Than Words* [8], which told the story of a child who was supposedly rendered autistic by vaccine. Both books were bestsellers in the autism community but their messages could not have been more different.

After my first book came out I heard from a number of autism organizations, the largest of which was Autism Speaks. They were newly founded, and already controversial when *Look Me in the Eye* was published. Their portrayal of autism was that of a monster that ruined marriages and stole children. While that played well for fundraising, it was challenged from the beginning by autistic people, who found that kind of talk offensive.

My own book and life story were about building my best life, just the way I was. Having learned through study that autism is a stable neurological difference, not subject to cure, I saw no other sensible course of action. When I read the narratives that were emerging I wondered how much proposed research would possibly benefit people like me. They were focused on finding a cause so they could find a cure. I saw that as totally irrelevant to an autistic person like me. My problems were how to get through school, how to find jobs, and how to sustain relationships.

As I met more autistic people I came to see how some seem far more impaired than me. I saw families where one person was autistic with no trace of autism elsewhere in the family tree. Other families seemed full of autistic people, in every generation. The cause of autism in my case and some others seemed evident—it was woven into our family tree. It wasn't so clear in some of the other families. That opened my eyes to the idea there may be many "autisms" and many paths into this thing we call autism.

Scientists began writing me as soon as my book was announced. They were eager to find autistic adults who could talk about their ideas for autism research. Those conversations led to my joining advisory boards at universities, at hospitals, and in government. It was there I began meeting autism scientists and policymakers.

In December of 2007, University of Washington child psychologist Geraldine Dawson was named Chief Science Officer of Autism Speaks. After reading my book she sought my input on the direction of autism science. I became the first autistic person to advise Autism Speaks on research to serve autistic people.

By that time I had visited a larger number of autism schools and programs, and talked to countless autistic individuals. One thing that came through very clearly was that we needed help with independent living. For some of us, that meant help with organization. For others we needed strategies to manage sensory sensitivities. Some needed help communicating with the non-autistic public around them.

As I began talking to autism researchers I realized I did not have to be a scientist to have a valuable perspective on autism research. My life as an autistic person allowed me to put proposed research in perspective with a key test: *What would this mean to someone like me?* All too often, proposed studies had no beneficial connection to actual autistic people.

That bothered me a lot, particularly as I learned about the breadth of medical problems afflicting autistic people. For example, epilepsy is managed among the non-autistic population but it's seemingly uncontrollable for many autistics. Many of us live with severe gastro-intestinal issues. Anxiety and depression are constant companions for most of us. Those should have been hot topics for research, but they were not.

When I looked at the research Autism Speaks was funding, I saw next to nothing that had potential to resolve the problems I saw among autistics. Instead they were heavily focused on basic genetics and biology. To me, the disconnect was obvious even as the researchers defended their current courses of action.

From the beginning, autistic people were skeptical of my involvement. Some asked why the science community would pay attention to me, a lone autistic who was not even a scientist. Others asked how I could have anything to do with a group that said such awful things about autistic people. The language Autism Speaks used to describe autism and autistic people was very troubling, but I believed they might change their message once exposed to actual autistic adults. Staffers like Dawson seemed to share that belief. In hindsight I see that thinking was naive. It was hard to imagine myself as diseased or damaged, but I understood those words made people open their wallets and I knew our community needed help.

My involvement was limited to recommending courses of research. I had nothing to do with Autism Speaks public statements. It always troubled me when people in the community thought I was an Autism Speaks employee or spokesperson, because I was never either of those things. It embarrassed me to be associated with them, but at the time they were the largest private funder of autism research in the USA and I thought my impact there might be more impactful than at a smaller organization.

I continued to advise Autism Speaks on science, and I also continued my service on autism committees at the National Institutes of Health (NIH), the Centers for Disease Control (CDC), and the Department of Defense. Of those groups, Autism Speaks was always the most controversial. I often wondered if it was worth staying involved, but I kept harboring a hope they would change their rhetoric.

Then in 2009 [9] Autism Speaks released its now infamous *I Am Autism* video which characterized autism as a monster, destroying lives and families. As an autistic person it was hard to see that video as anything but a demonization of the essence of what I am. Thousands agreed, and the ensuing public relations debacle highlighted the widening gulf between certain parents and the emergent community of autistic adults. Parents who imagined themselves as victims of the "autism monster" justified themselves by saying autism in people like me was somehow different, even as the evolving science said that wasn't true.

There was little doubt that I was less disabled than some other autistics, but people vary in every community, and one thing we autistics tend to agree on is our perception of ourselves: we tend to feel less disabled than outside observers judge us to be. I've met a few autistics who think autism is a horrible disability and want a cure, but most of us accept that we are what we are, and do the best we can.

I tried very hard to deliver this message to parents inside and outside of Autism Speaks. In addition to being an autistic person, I was a parent of an autistic son, and I thought I understood how they felt, even if I did think some of their ideas were unhealthy and counterproductive.

In July of 2012, I attended a strategic planning meeting for Autism Speaks in San Francisco. I sat in a room with a dozen esteemed scientists from some of the most prestigious institutions in the world. At that meeting, I proposed that we ask the Autism Speaks governing board to drop the word "cure" from its mission statement. The scientists were all in agreement. Instead, we proposed that we funded research to understand the biological basis of autism and how we might relieve specific aspects of autistic disability. The scientists agreed that no one was researching the broad idea of "cure," and evidence suggested "cure" was not a realistic goal. Remediation of disability is a realistic goal.

After the meeting I expected to hear something from Geri Dawson. Perhaps they'd want me to attend their board meeting, to explain this proposed mission change. The main Autism Speaks board meeting came and went, and nothing happened. I called Geri. "They were not ready to hear that," she told me. The resolution had never even been discussed. I was deeply disappointed, and sad.

A few months later Geri left for a new job running the autism center at Duke University. I was sorry to see her go, but I understood as I shared her frustration with the group's fixation on what I believe we both saw as unhealthy ideas. Rob Ring—a former pharmaceutical executive—was named to take her place.

In October of that year (2013), Autism Speaks announced an autism summit to be held in Washington, DC. Significantly, there was not a single autistic person scheduled for attendance. Then in early November Autism Speaks founder Suzanne Wright followed that up with a truly horrific op-ed. In it, she suggested that millions of autistic people were "lost," taken from society by the monster autism. She said families, and people like me, were "barely living." The response from autistic people was predictable.

I found her article extremely offensive. It made me think of the *I Am Autism* piece they had published four years previously. Worst of all, people in the autism community blamed me for being complicit in the newest Autism Speaks debacle. More than a hundred people wrote me to ask how I could be associated with an organization that promulgated ideas like Wright's.

That was a question I could not answer, because I felt the same way. In addition, I felt a deep sadness, realizing my four years of advocacy work within the organization had not made one bit of difference to the Wrights, who headed the organization. I wrote a letter to Liz Feld, the president.

In my letter, I said:

> Autism Speaks is never going to be accepted by the broader community of autistic people if they continue the fear-mongering and "sick child" talk…This kind of talk does not do any of us any good.

> The idea that [Mrs. Wright] would once again convene a "summit" without any meaningful autistic representation is extremely troubling to me, particularly because we've covered this issue before.

> I'm starting to feel like Mrs. Wright is in a very different place than most of the people I see in the autism community…Is Autism Speaks going to

be able to shift its focus away from her "diseased child" model to focus on consulting with autistic people of all ages…about how their needs might best be served, in a non fear driven environment?

After two days without an answer, I made my decision. On November 13, 2013, on my way to deliver an autism talk in Grand Rapids, Michigan, I wrote a resignation letter. I sent it to Liz Feld's email and I also posted it on my blog (http://jerobison.blogspot.com). The reaction from the autism community was swift.

Most commenters supported my decision to stand up for my beliefs. Online comments (there were over 200 on the blog post alone) were critical of the ideas expressed by Mrs. Wright and of Autism Speaks for continuing to give her views a home.

There were others, though, who expressed solidarity with Mrs. Wright's ideas. Reading their comments, most seemed to be parents who blamed autism for stealing their own children. While I understood how parents might feel that way, I had spoken about how unhealthy that kind of thinking was, both for the families and for the autistic children. Yet it persists in a portion of the community to this day.

I hoped my departure would precipitate some kind of action, but Autism Speaks remained silent on that issue. A number of staffers spoke to me privately, expressing sadness or regret over my decision to leave. They all seemed to understand how I felt. In the past five years many have moved on to other jobs.

Autism Speaks' research portfolio remains heavily weighted toward biology and genetics, and studies that are unlikely to materially benefit this generation of autistic people. I've come to see disagreements with this view as illustrative of the disconnect between what we autistics say we want and need, and what researchers say they should study to help us. It's one reason we autistics need more influence over the research agenda.

The organization's silence led me to consider whether I overestimated my importance to the group. Geri Dawson and the scientists were sincere in seeking my input and I feel we learned much from each other. Yet theCed Wrights seemed to exist in a separate world, and I don't know if they even knew who I was. I'd been the only autistic person to have a voice in science, but that did not seem to make a lasting difference. I'd spoken often, but I wasn't heard by the people in power. Meanwhile, I gained a false confidence as lower-level staffers agreed with my positions. None of us had any influence with the leaders. As time passed and I reflected, I realized Autism Speaks was misnamed. They do not speak for autism, or autistic people, and they never did.

Their founder was a media executive; their name a marketer's creation. Many of the staff described themselves as non-profit professionals, and none of the senior people were autistic. They were very effective at fundraising, and painting a picture of autism that elicited widespread sympathy.

Autism-as-tragedy helped them raise hundreds of millions of dollars. Groups like Charity Watch reported that they spent lavishly on themselves and their organization compared to other medical nonprofits. Their annual reports told a sad story. Monies raised locally paid headquarter's salaries and supported distant researchers. Very little returned to the communities who raised the funds. Perhaps it's all about the money, I thought, and my ideas of acceptance and fitting in are not a basis for tens of millions in donations the way "stolen children" and "ruined families" are.

I had imagined I was making a difference on their science board, but the Wrights called the shots when it came to investing the organization's money, and the research I had advocated for took a back seat to the Wright's agenda, which appeared to be biology and cure. Had I been able, I would have made different choices.

Autism Speaks would probably disagree with me, but I felt then and feel today that their focus on causes and cures did very little to help the millions living with the reality of autism. From the beginning of my autism advocacy, I have kept that goal in sharp focus and I'm quickly frustrated when others can't do the same.

I met a number of bright dedicated researchers while volunteering for the Autism Speaks science board. Many of us continue to work together today, on government boards, with the World Health Organization (WHO), and with the International Society for Autism Research (INSAR), the professional society for autism researchers. Those other organizations have changed significantly in response to autistic input. We have a strong voice in creation of our government's autism plans. The WHO's International Classification of Functioning, Disability and Health (ICF) Autism Core Set recognizes both disability and exceptionality in us, thanks to autistic input (http://jerobison.blogspot.com/2018/04/autism-ability-disability-and-icf-core.html). INSAR has encouraged making autistic people research collaborators. American public health agencies do the same. They expand their embrace of the community every year.

Since my departure Autism Speaks has seen considerable upheaval. Liz Feld left, followed by Rob Ring. Mrs. Wright passed away in 2016 and her husband resigned his position a short while later. The organization's funding of research has dropped these past few years.

Two years after my resignation Autism Speaks announced Stephen Shore and Valerie Paradiz were joining the organization's board. They were actual autistic people with a say in the group's governance. While I applaud them for doing that, I'm still waiting for substantive autistic-led initiatives from the organization.

Elsewhere in the autism community, there is an evolving and sometimes heated dialogue about who should speak for autistic people. For the last few decades advocacy has been the province of parents, grandparents, and professionals of various stripes. They were the ones who rose up and demanded services in response to the wave of new diagnoses beginning in the 1990s.

Today the kids they advocated for have grown up, and many are finding their voice. One principal venue for autism advocacy is now the Interagency Autism Coordinating Committee (IACC), which produces the strategic plan for autism for the US government. IACC guides NIH, CDC, Defense and other government agencies as well as private groups. At first IACC's advisers were autism researchers, clinicians, and parents. For the past decade, IACC has also had actual autistic people as members. I was appointed in 2011 and continue to serve as of this writing.

At IACC meetings I routinely see differences of opinion between non-autistic parents and actual autistics. The conversation often turns to who should have the primary voice, and I see that it's very hard for non-autistic parents to let go. Yet I feel that is what must happen. Adults speak for themselves in all other walks of life. Autism is a lifelong condition, not a childhood disorder (as was once thought). Autism research, therapy, and policy should be guided principally by autistic adults. It's that simple.

Elsewhere in society we accept the idea that anyone who speaks for a group should be a member of the group. By that reasoning any spokesperson for autistic people should be autistic. A parent can certainly speak for autism parents, but that is a different community and like all parents, their wants and needs are sometimes at odds with those of their children, particularly as the children grow up.

The day may come that Autism Speaks is led by actual autistic people. I hope that happens. Alternately, Autism Speaks may remain primarily a parent advocacy group for families with young children on the autism spectrum. That is effectively what they were in the beginning, and where they may be most at home.

Actual autistic people seem more drawn to autistic-founded and autistic-led groups like the Autistic Self Advocacy Network (ASAN) and I expect they will grow more powerful as their membership grows and ages.

I joined the Autism Speaks science board in the hope I could help move their science in a direction that would be more beneficial to autistic people. At the time I thought their legacy would be good autism science. I left the Autism Speaks science board because of their hurtful depictions of autism and autistic people. Autism Speaks did not make a huge mark in science, and with drops in funding their significance in that world has diminished. It's toxic rhetoric that has become the organization's legacy.

Meanwhile we autistic people are still here. We're not missing, and we're not lost. Monsters will not take us, because we are strong. When it comes to policy, parents and clinicians certainly have a say, and deserve a seat at the table, but the table rightly belongs to us. We are autistic people.

References

1. Attwood, T. (1997). *Asperger syndrome: A guide for parents and professionals.* London, UK: Jessica Kingsley Publishers.
2. Lord, C., Risi, S., Lambrecht, L., Cook, E. H., Jr., Leventhal, B. L., DiLavore, P. C., et al. (2000). The autism diagnostic observation schedule-generic: A standard measure of social and communicative deficits associated with the

spectrum of autism. *Journal of Autism and Developmental Disorders, 30*(3), 205–223.
3. Robison, J. E. (2007). *Look me in the eye: My life with Asperger's*. New York, NY: Crown/Archetype.
4. Tammet, D. (2007). *Born on a blue day*. London, UK: Hodder.
5. Grandin, T., & Scariano, M. M. (1986). *Emergence: Labelled autistic*. Novato, CA: Arena Press.
6. Williams, D. (1992). *Nobody, nowhere: The extraordinary autobiography of an autistic*. New York: Times Books.
7. Shore, S. M. (2003). *Beyond the wall: Personal experiences with autism and Asperger syndrome*. Shawnee Mission, KS: AAPC Publishing.
8. McCarthy, J. (2007). *Louder than words: A mother's journey in healing autism*. New York, NY: Plume.
9. Autism Speaks. (2009). *I am autism*. Retrieved March 1, 2019 from https://www.youtube.com/watch?v=9UgLnWJFGHQ&t=14s.

Open Access This chapter is licensed under the terms of the Creative Commons Attribution 4.0 International License (http://creativecommons.org/licenses/by/4.0/), which permits use, sharing, adaptation, distribution and reproduction in any medium or format, as long as you give appropriate credit to the original author(s) and the source, provide a link to the Creative Commons license and indicate if changes were made.

The images or other third party material in this chapter are included in the chapter's Creative Commons license, unless indicated otherwise in a credit line to the material. If material is not included in the chapter's Creative Commons license and your intended use is not permitted by statutory regulation or exceeds the permitted use, you will need to obtain permission directly from the copyright holder.

17

Covering the Politics of Neurodiversity: And Myself

Eric M. Garcia

I first learned about neurodiversity in the summer of 2015 as a reporter for *National Journal*. I was recruited there largely by Ron Fournier, a columnist I befriended when I emailed him as a college student after he wrote a story about his son with Asperger's syndrome. Ron suggested I read a book called *Neurotribes: The Legacy of Autism and the Future of Neurodiversity* by his friend Steve Silberman [1]. It was the first time I read about the neurodiversity movement.

At around that time, I was writing an essay that mixed reporting with my own personal experiences as an autistic reporter for one of the last print editions of *National Journal* [2]. Even though I had been diagnosed with what was then Asperger's syndrome when I was kid, I had almost no experience writing about my own experiences or exploring much about the autistic community in D.C. or at large. I knew vaguely what autism was but did not know how my own story fits into the larger context about autism and certainly didn't understand what neurodiversity meant.

E. M. Garcia (✉)
Washington, DC, USA

I hoped it would be a fun, chatty piece about the secret lives of autistic reporters and people who work in politics in a city that values social capital even when autism makes socializing difficult.

But the magazine's editor, Richard Just, asked me why this story should exist. In a mix of frustration and hubris, I said American society focuses too much on "curing" autistic people and not enough on helping autistic people, particularly adults. Richard said he wanted 10,000 words on it.

The essay—a slim 6500 words—set my trajectory today [2]. While I didn't have a clear idea of what the term *neurodiversity* meant, the piece argued policy should bend toward this radical idea. The notion was society should accept and accommodate people with autism, dyspraxia, ADHD, or other conditions considered an abnormality. Neurodiversity wasn't diminishing the specific challenges of autistic people but rather, the essay argued, like other disability rights movements before it, society should welcome neurodivergent people and give them the tools necessary to live a life of dignity. Through research, interviewing, and reporting, I see that neurodiversity is an argument for civil rights; that instead of the world trying to make us be more "neurotypical," the world should celebrate our atypicality and accommodate accordingly.

As a reporter, when I am ignorant about something, I call people knowledgeable about it. But when it came to reporting about autism, I noticed while many journalists quoted parents of autistic children, professors, nonprofits, or legislators, they often ignored autistic people.

Though they have a gross oversight, I don't blame most journalists. Oftentimes when we are ignorant about a new beat, we type in an assortment of words into Google and hope for the best. With autism, the first results are often groups with little autistic representation like Autism Speaks—which only recently began adding autistic people to leadership positions [3]—professors, and parent leaders. Similarly, because autistic people were seen as a burden, they went unheard while their parents' struggles took center stage. Hence, most autism stories don't feature the people with the most expertise: the people who live with autism on a daily basis.

That matters because those who are most heard get the most attention and in turn, are the ones who shape policy. It is for this reason I made it a point when I wrote my initial story to make sure I interviewed as many autistic people as possible. It was largely through Dylan Matthews, a writer at Vox.com, that I was introduced by groups like the Autistic Self Advocacy Network; its leaders Ari Ne'eman and Julia Bascom; as well as other prominent autistic advocates like John Elder Robison, Lydia Brown, Liane Holliday Willey, and Dena Gassner.

I became more well-acquainted with other autistic and neurodiversity advocates like Finn Gardiner, Sara Luterman, Samantha Crane, Morénike Giwa Onaiwu and Sharon daVanport. All of them were essential to understanding permutations of autism and what it meant to celebrate neurodiversity, and in turn they gave me resources fellow reporters didn't have: a cache of autistic people whom I could consult anytime autism popped in the news.

It might sound trite to say something "changed my life" but writing and publishing that essay completely upended my career. Since I began as a college reporter, I had been taught to avoid making yourself the news and to be an objective arbiter of the news. But I remember the day the essay being published and receiving a call from Ron on what would be my last day at *National Journal* with him in tears simply saying "holy shit" with pride at how moved he was by it.

That same day, I got a call from Sen. Orrin Hatch, then the most senior Republican in the United States Senate who was then chairman of the Senate Finance Committee, saying that he had helped sponsor the Americans with Disabilities Act that passed in 1990 and that if I ever wanted to speak with him about autism, I was welcome to visit his office. I had asked Sen. Hatch questions multiple times in the halls of Capitol Hill and was never any more than a nondescript nameless reporter. But now, I was someone whom he knew on a first-name basis and could pick out.

Similarly, for a week, when I opened Twitter I would see journalists whom I deeply admired write about how my essay would "move you." Some of them were bylines I read regularly in *The Washington Post*, *The New York Times*, and people elsewhere, and whom I regularly emulated. A few months after the publication of that essay, I saw a journalist I admired and emulated while I was in college at a gala and he told me the piece was amazing. While I was humbled by him complimenting me, in the back of my mind I remembered when he didn't return my email when I was a lowly intern in Washington asking if we could meet for coffee.

At first this success gave me a sense of vindication about my decision to write. I could now pitch stories about autism and people would consider me credible because I poured my heart into this piece. It also landed me a book contract that would further burnish my credentials as an "autism expert" journalist.

But in the same respect, while I wanted my piece to shed light on autism, I feared that I was just becoming another "inspirational story" about how I overcame autism. I worried that in my attempt to counter the narrative that autism is a debilitating condition, I somehow inadvertently lifted myself up as above being autistic. I don't think I ever said it to myself but possibly subconsciously, what I decided to do as a journalist was to show that my self-portrait was not the outlier but was in fact common among autistic people and that we were in fact, deserving not of pity but of shared humanity.

What was peculiar about this newfound attention was that the essay was published on my very last day at *National Journal* as I was leaving to start a new job as a staff writer at *Roll Call*. To boot, I felt like that was the best way to be an effective voice for autistic people; I didn't want to continue dining out on that essay and be seen as "the autistic reporter." I felt—and continue to believe—the best way that I could assist autistic people was to show I was as good of a reporter as my neurotypical counterparts.

As a result, I made it a point that when I began writing my book, which is still in progress as of February 2019, that I would not just make it into a personal memoir, but rather that I would rather chronicle the lives of other autistic people and show how society often creates more obstacles for autistic people than the actual autism does. Similarly, when I was asked to write a piece for *the Washington Post* about how parents teach autistic boys about consent in sex and relationships, I chose to flip it on its head and say that it is important to teach autistic people consent because they deserve to have as fulfilling love lives and sex lives—if they so choose—as neurotypical people [4].

The more I researched and the more I saw how politicians discussed autism, the more I felt that while I did not want to become part of the news and did not want to make myself the story, my own understanding of autism and my connections to the disability community could allow me to contextualize autism and bring new voices into the public sphere that would not otherwise have been heard.

Incidentally, one of my first opportunities to bring this perspective to my reporting came within months of me starting my job at Roll Call when Democratic nominee Hillary Clinton announced a comprehensive platform on autism [5].

When Clinton embarked on her second presidential campaign in 2016, she released a comprehensive set of policies that read like a neurodiversity wish list like conducting a nationwide survey of autism prevalence in adults, banning physical and chemical restraints, and creating an Autism Works Initiative [5]. As someone who had just written an essay about this a month ago, much of what I read in her policy platform read like things I had heard advocates dream would happen in our interviews.

Clinton's own maturation on autism was reflective of how the politics of autism changed. In a 2007 speech in Sioux City, Iowa Clinton spoke about being a student at Yale Law School and spending a year at Yale's Child Study Center in the 1970s, when autism was still largely blamed on unloving mothers and told a story about a female friend in Little Rock, Arkansas whose son had autism.

"And I spent time in her home, I spent time with her and her son and my instinct perhaps as a mother was that this could not be the explanation," she said [6].

As a senator, Clinton introduced her own legislation called the Expanding the Promise for Individuals with Autism Act, a bill that never passed but was introduced months before her Sioux City speech [7]. In the Sioux City speech, she talked about helping train educators to handle autistic students and providing proper services for autistic adults, which is still groundbreaking by today's standards.

"With access to the right types of services, including housing, vocational rehabilitation, we can help adults with autism live rich and full lives," she said in the speech [8].

But even with these important steps toward progress, Clinton's speech was still couched in the political zeitgeist about autism at the time. In the sentence before saying there are insufficient services, Clinton said "we don't know how to cure it and we don't even know the best ways to treat it" [8].

Similarly, Clinton credited parents when she said, "driven by their love and devotion, mothers and fathers across the country have raised awareness, demanded funding, and opened our eyes to the needs of so many children" [8].

Clinton's longstanding focus dating back to at least her days in law school shows her shift in rhetoric from her first to her second presidential run was not solely for political expediency but which came from listening to autistic people.

Part of this was fueled by the fact that Clinton had the support of autistic self-advocates like Ne'eman, who was on the conference call for reporters, including myself, to announce the roll-out of these policies. He said at the time "the fact it was requested and the fact many of these priorities come directly from the community is extremely significant" [9]. While there were a number of estimable reporters on the line, the fact I am autistic I felt could give me an advantage. While other reporters wrote a simple breakdown of the policies, I was able to call autistic self-advocates, who would be the most affected by these policies and chronicle their varied reactions. Being autistic and having that institutional knowledge gave me a roadmap other reporters didn't necessarily have, without bringing my own narrative into the story. It showed that I knew which perspectives mattered the most with these things.

Furthermore, my understanding of autism allowed me to debunk hoaxes and pseudoscience that arose in the campaign. Then-candidate Donald Trump decried autism as being "an epidemic" during the primary debates and told a dubious anecdote about a friend's child becoming autistic after vaccinations [10]. Similarly, during the campaign, Trump met with Andrew Wakefield, the discredited former doctor responsible for promoting the bunk theories of vaccines-autism causation, and Gary Kompothecras, another major promoter of the anti-vaccine theories [11].

After his election, Trump continued his egregious peddling of vaccine theories by promising Robert Kennedy Jr., a prominent anti-vaccine activist, that he would chair a vaccine safety commission [12].

Thankfully, it appears that Trump has not mentioned the idea frequently and Kennedy told *the Guardian* that the administration "cut off all communication with people who care about this issue" [13].

But despite backing away from vaccines, Trump has continued to peddle harmful narratives. On April 2, 2017, Trump's presidential proclamation for World Autism Day read "My Administration is committed to promoting greater knowledge of [autism spectrum disorders] and encouraging innovation that will lead to new treatments and cures for autism" [14].

Similarly, Trump lit the White House blue, which is emblematic of Autism Speaks, for Autism Awareness Day, an action that was criticized because it sees autism as a problem to be fixed and isn't driven by autistic people themselves [15]. Unlike Clinton's comments, these came after better understanding about autism as a condition and after autistic voices have been heard in the public square.

As an autistic political reporter, this new era where there is a resurgence of anti-vaccine conspiracy theories and autism is still used as an epithet at times, often makes me question my place in the autistic community. I am not and will never claim to be the sole voice for autistic people in political media. My other autistic peers are more than capable and probably better at writing about the topic than I am. Similarly, as someone who is now an editor at *The Hill*, another congressional trade publication, I feel an obligation not to pick one side. But it can at times feel maddening when I feel like the side of autistic people doesn't even get a chance to say anything, even those of my autistic brethren who can't verbally speak.

I don't really know if I have changed people's hearts and minds about autism. At times I worry that I have an inflated view of myself as the guy who is trying to monopolize "the autism beat" in the Washington press corps. But for the time being, I see so few figures who have the context, the understanding and frankly, who care enough to provide those things to news consumers. How could I, when I speak and regularly interact with autistic people, not want to ensure others get our narrative correct? I hope that by both my presence among my colleagues and peers and by the words I write, I can deliver truths not by protesting or lobbying for rights, but by changing who people regard as worthy to deliver their news and ensuring that autistic people are accurately portrayed and their needs be seen as legitimate.

References

1. Silberman, S. (2015). *Neurotribes: The legacy of autism and the future of neurodiversity*. New York, US: Penguin Books.
2. Garcia, E. (2015, December 4). I'm not broken. *The Atlantic*. Retrieved from https://www.theatlantic.com.
3. Autism Speaks. (2015, December 15). *Autism Speaks welcomes three new board members* (Press release). Retrieved March 5, 2018, from https://www.autismspeaks.org/news/news-item/autism-speaks-welcomes-three-new-board-members.
4. Garcia, E. (2017, April 27). Autistic men don't always understand consent. We need to teach them. *The Washington Post*. Retrieved from https://www.washingtonpost.com.
5. Clinton, H. (2016, January 6). *Autism policy*. Retrieved March 4, 2018, from https://www.hillaryclinton.com/issues/autism.
6. Clinton, H. (2007, November 24). *Remarks at the autism event with Sally Pederson in Sioux City, Lowa* (Transcript). Retrieved from https://www.cs.cmu.edu/~ark/CLIP/candidates/clinton_h/2007.11.24.remarks_at_the_autism_event_with_sally_pederson_in_sioux_city_iowa-overlay.html.
7. Clinton, H. (2007, March 20). *Clinton introduced autism bill to promote services for those affected by autism* (Press release). Retrieved from https://www.presidency.ucsb.edu/node/297147.
8. Lorentzen, A. (2007, November 25). Clinton would boost autism funding. *The Associated Press*. Retrieved from http://www.washingtonpost.com/wp-dyn/content/article/2007/11/24/AR2007112400924.html.
9. Garcia, E. (2016, January 6). Autism advocates cautiously optimistic on Clinton proposal. *Roll Call*. Retrieved from https://www.rollcall.com.
10. Beckwith, R. T. (2015, September 18). Transcript: Read the full text of the second republican debate. *Time Magazine*. Retrieved from http://time.com.
11. Kopplin, Z. (2016, November 18). Trump met with prominent anti-vaccine activists during campaign. *Science*. Retrieved from http://www.sciencemag.org.
12. Phillip, A., Sun, L. H., & Bernstein, L. (2017, January 10). Vaccine skeptic Robert Kennedy Jr. says Trump asked him to lead commission on "vaccine safety." *The Washington Post*. Retrieved from https://www.washingtonpost.com.

13. Smith, D. (2018, February 21). Trump appears to abandon vaccine sceptic group denounced by scientists. *The Guardian*. Retrieved from https://www.theguardian.com.
14. Trump, D. J. (2017, March 31). *President Donald J. Trump proclaims April 2, 2017 as World Autism Awareness Day*. Retrieved March 5, 2018, from https://www.whitehouse.gov/presidential-actions/president-donald-j-trump-proclaims-april-2-2017-world-autism-awareness-day/.
15. DeJean, A. (2017, April 17). The White House turned blue for "autism awareness." That's actually bad for autistic people. *Mother Jones*. Retrieved from https://www.motherjones.com.

Open Access This chapter is licensed under the terms of the Creative Commons Attribution 4.0 International License (http://creativecommons.org/licenses/by/4.0/), which permits use, sharing, adaptation, distribution and reproduction in any medium or format, as long as you give appropriate credit to the original author(s) and the source, provide a link to the Creative Commons license and indicate if changes were made.

The images or other third party material in this chapter are included in the chapter's Creative Commons license, unless indicated otherwise in a credit line to the material. If material is not included in the chapter's Creative Commons license and your intended use is not permitted by statutory regulation or exceeds the permitted use, you will need to obtain permission directly from the copyright holder.

18

"A Dream Deferred" No Longer: Backstory of the First Autism and Race Anthology

Morénike Giwa Onaiwu

The way I perceive things, my involvement as one of the editors of the first-ever anthology on race and autism, *All the Weight of Our Dreams: On Living Racialized Autism* [1], was by pure accident.

An extremely fortunate accident that I will eternally be grateful for, but an accident nonetheless.

The anthology itself was conceived in the mind of Lydia X. Z. Brown, an Autistic disability rights activist with whom I serve on the Board of Directors of the Autistic Women and Nonbinary Network (AWN Network, a grassroots Autistic-led nonprofit for gender minorities that was founded in 2009), and AWN Network enthusiastically agreed to partner with Lydia to bring the project, announced in summer 2014, to fruition.

Having learned about the idea early in its development, I, like many others, embraced and helped to promote it. Community support for the anthology was vital for a variety of reasons—fundraising being one of them. Though AWN Network had successfully secured a small amount

M. Giwa Onaiwu (✉)
Houston, TX, USA

of grant funding and sponsorships for the anthology, in order for it to become a reality a lot more money was going to be needed…and as none of us had much money, it was clear that we were going to have to rely upon crowdfunding to obtain the rest. However, to get people to donate, we knew it was crucial for us to make a convincing case for why this anthology needed to be published. So in the beginning, when the anthology was just a creative idea, a lot of the focus was on finding the means to make it happen. To justify a reason for its existence.

Ultimately, the most compelling reason we could come up with, and the one that successfully enabled us to reach our $10,000 fundraising goal in approximately one year ($6712 from 143 different supporters, $2000 from a state Developmental Disabilities Council, $1000 from the Autistic Self Advocacy Network, and the rest from in-kind donations) was simply the truth. And that truth is one that Autistic people of color like myself, Lydia, E. Ashkenazy (a Deaf multiracial Autistic woman who is also one of the editors of the anthology), and the more than 60 different individuals who would later become contributors to our anthology share: that the existing public discourse about autism is glaringly incomplete.

The experiences, stories, and images of people of color on the autism spectrum are conspicuously absent both in the public view and sadly even within the sphere of disability. The glaring lack of visibility of people of color within the disability community which prompted Vilissa Thompson, a Black disabled writer and social worker, to coin a phrase that became a viral hashtag (#DisabilityTooWhite) clearly exists within the autism realm as well. In fact, it might be an even greater problem.

Sharing that simple yet profound truth over and over was successful. Our plight resonated with people and they supported us both in word and in coin. And eventually there was a sufficient amount of money to begin the actual work of creating the anthology.

Initially, the plan was for Lydia to serve as the Lead Editor of the project and E. Ashkenazy as Project Manager. However, as a result of the overwhelming response to the call for submissions (submissions were pouring in from various countries across the globe), it became apparent that another member of the primary editorial team would be needed and I was selected. So although I had not originally planned to be involved with the anthology beyond providing general external support for the idea, things changed, and when approached to join the team I happily accepted. At the time I was not fully aware of the depth of the task we had before us nor could I grasp its overall significance; I was just excited, but it would later dawn on me.

We were quite an eclectic team, and we had a lot to learn. We were all Autistic. We were all people of color: Lydia is East Asian; E. is multiracial; I am Black. All of us loved to write and had even had some previous work published, but none of us considered ourselves professional writers. None of us had an educational nor work background in journalism; none of us had ever worked for a magazine, book publisher, or newspaper. None of us had ever been involved in a project quite like this. In addition, we all had major time constraints due to our busy lives and numerous responsibilities. Lydia was enrolled in law school at the time; E. was running a professional skating and dance company; I was teaching. E. and I had eight kids between the two of us. All three of us had various advocacy commitments we were engaged in that required time and energy, and we all had different work schedules. Furthermore, we all lived in different parts of the US: E. lived on the West Coast, Lydia on the East Coast, and I was in the South.

However, all three of us were devoted to the project and willing to "learn on the job" and challenge ourselves to make things work. We all possessed a strong work ethic, an enormous amount of enthusiasm, the ability to "think outside the box" which helped us to grasp new things quickly, and the ability to hyperfocus. We had different working styles: one of us was a planner who liked to make lists and preferred to stay ahead of deadlines; another was a perfectionist and procrastinator who would have periods of inactivity/dormancy and then become infused with a creative burst of energy that enabled rapid completion of a voluminous amount of work at the very last moment; the other person was a combination of these two types.

Our differences worked in our favor as we all brought unique strengths to the team and were able to support one another well if we faltered because we didn't all share the same weaknesses. We were united in our belief in the importance of the anthology and in our respect for one another. Over the years we maintained regular (sometimes multiple times daily) communication via email, message, and other electronic means. Together, along with several other dedicated individuals, we built something real, something monumental, something beautiful.

From the very beginning, it was important to us that the anthology remained true to Lydia's original vision (which was one that we shared). It was also important that we created something that we believed in and could be proud of. We realized early on that many established "best practices" that are utilized in the development of publishing a book of this volume, while generally helpful and useful for others, were not going to necessarily work for us. Our process was going to look and feel different in several ways, and was going to be both non-linear and nontraditional. Because of this, we weren't going to have a formal framework, "blueprint", or set of instructions that we would be able to rely on. Where there was no precedent, we had to be ready to trust one another and forge our own way—and we did.

We developed and agreed upon several guiding principles throughout the process of birthing the anthology. Some of them include:

Nothing About Us Without Us. Every member of the (lead) editorial team, which consisted of the three of us, had to be an Autistic person of color (PoC), and we needed to have full autonomy over our work.

We did allow Autistic people who were not PoC to assist with the project itself, and many did, including Shain Neumeier and Clare Barber-Stetson (White Autistic colleagues who assisted with reviewing submissions and providing feedback on any necessary revisions), Amanda Gaul (White Autistic colleague who provided legal and financial assistance), Melanie Yergeau (White Autistic colleague who completed most of the formatting [along with Tracy Garza who is an Autistic PoC]), and Lori Berkowitz (White Autistic colleague who provided web and technical support). We also had a number of White allies in the community help with generating interest and media attention for the anthology. Solidarity is important, and we appreciate the efforts made by the many people who supported this project.

However, it needed to be **our** project. "Our" meaning Autistic PoC. Led by us, coordinated by us. Not by White people—no matter how supportive and like-minded they might be. It was imperative that we did not perpetuate the false notion that people who do not have the lived experience of being a PoC are somehow able to effectively represent us.

Every key decision about the anthology had to be made by and approved by us—period.

In that vein, we strived to have as many Autistic people, and especially Autistic PoC, involved in various aspects of the project as possible. The only contributors we accepted for publication were from among those who identified as Autistic PoC. We also intentionally sought out a graphic designer who was an Autistic PoC (Finn Gardner) for the cover art of the anthology. Although not a formal part of the anthology, another Autistic PoC, Sharon daVanport (AWN Executive Director), provided logistical support throughout the entire process. Moreover, we set aside funds to hire an Autistic PoC after the anthology was published to assist with a social media and marketing campaign.

We welcomed, and still do welcome, any and every person to read the anthology—regardless of what race they are and whether they are Autistic or not. It is not something that is only FOR us. But it had to be only BY us—and only us, and that was non-negotiable.

Everybody Gets Paid was another guiding principle that we felt strongly about and one that we will never compromise on.

Far too often the labor of marginalized people is minimized and treated as if it were of little value; meanwhile, those with vastly more privilege go on to profit enormously from the resources they have derived from those same seemingly "less valuable" people. Academics and professionals amass a plethora of publications, enhanced research portfolios, and career advancements on the backs of stakeholders whose only acknowledgment is a gift card and a contrived one line "thank you" written in small print at the end of their abstract or on their last presentation slide (or begrudgingly acknowledged in public decades later after numerous lawsuits, as in the case of the family of Henrietta Lacks).

We refused to be like that. We were not going to ask, nor expect Autistic PoC to contribute to an anthology and not get fair market value for their work. Marginalized people are frequently expected to possess a sense of altruism that is not expected of more privileged people; we are guilted into doing hordes of unpaid work to "help" our people, and/or told that we are getting "exposure" in lieu of paid wages. Well, you can't eat exposure. Your landlord won't accept exposure in place of a rent payment. Exposure doesn't buy diapers and milk. Meanwhile, others are paid for their time and effort. We respect and value the work of our people and if they could not get paid then we were not going to go through with the anthology. We researched market rates to determine what would be fair compensation and we made sure that every single person—from the contributors to the formatters to the attorney to the graphic designer, etc.—received payment for their involvement with the anthology.

We were warned by several external parties that this was not "sound business practice" and that our decision to do this was not recommended. We were told that this would not be profitable. We chose to reject their warnings as we believe in people over profit. Knowing the high rate of unemployment and underemployment among Autistic adults and the many challenges our community faces, how could we even conceive of taking advantage of our own people and cheating them out of what they deserve?

You Define You was a third guiding principle. Race is complicated and messy. So is neurology. Although our requirement that **all** contributors whose work was accepted for publication had to be Autistic PoC, we were adamantly against "policing" people's identities with regard to race and being Autistic. Phenotype does not always equal genotype; we refused to define race in a narrow, binary manner. ALL Autistic PoC, including biracial, multiracial and "White passing but PoC identified" individuals were welcome.

Additionally, Autistic PoC from any part of the spectrum were included, regardless of whether a person had a "PDD-NOS" diagnosis, an "Autism Spectrum Disorder" diagnosis, whether the person had self-diagnosed, etc. We did not require any of our contributors to have to conjure up "proof" that they were enough of this category or that category. You define YOU. We recognize that many barriers exist for Autistic PoC to obtain formal diagnoses should they choose to do so and also acknowledge that for some of our people it is neither advisable nor safe to do so.

For the anthology, we opted to accept the validity of people's self-identification as stated.

It's Not a Term Paper was another guiding principle of ours, and one that we have found to be somewhat controversial. We made the decision that we were not going to edit our contributors' pieces to death. Too often rigorous, elitist Western standards are applied to PoC, especially disabled ones, when they are inappropriate for the context. We felt that this would have been an instance where "standard" editing procedures would have created more harm than good. Autism is, among other things, a social communication disability; it is to be expected that there may and likely are some differences in how things are communicated—and that's without taking cultural factors (because all of the contributors are PoC) into account.

We preferred to prioritize retaining the true essence of what our contributors were trying to say over chopping up and "white-washing" their words for the purposes of ensuring they had perfect grammar, punctuation, and spelling. Especially since the accepted pieces come from contributors from a range of ages (one as young as five years old!), countries (including countries where English is not a commonly spoken language), backgrounds, and levels of schooling (from very little formal education to graduate school). Unless there was a pressing need to make grammatical corrections, we chose for the most part to capture our contributors' intended meaning AND original wording as is.

This guiding principle applied to us as well, not just to the contributors; if you review my own preface to the anthology carefully, you might see a word repeated twice accidentally. Other places in the anthology you might find various colloquialisms, ethnic slang, misspellings, etc. Yes, we saw them all—and yes, we left them there. Humans do not write perfectly; why pretend that we do? Why create a written version of a heavily "PhotoShopped" image that barely looks like the real thing when we have an opportunity to be authentic in our presentation? For us, there was no valid reason why and were plenty of reasons not to.

Real Transparency was another guiding principle of ours in putting together the anthology. We viewed this project as one that belonged to the community. For before the anthology even had a name, before we had even a single contributor, when it was just a dream floating around in the cosmos that we were trying to make tangible…our community backed us. It was they who provided the bulk of the money for us to pay everyone involved. It was they who encouraged us when we hit snags in the process (which happened more than we had anticipated, unfortunately). It was they who shared with their friends, family, and coworkers about the anthology that was underway to help us to get buyers.

On our AutismAndRace.com website as well as on social media and in other places we were open about timelines that we struggled with (not surprising given that this was a new undertaking for all of us, we are all Autistic people who struggle with executive functioning, and that we were juggling this project among a number of other responsibilities). We were forthright about delays, but also about successes and exciting developments that occurred along the way. We also regularly expressed our gratitude for the support and encouragement we were receiving. Good or bad, promising or disappointing, pretty or ugly...we kept it 100% real with our followers the whole way through.

We've Only Just Begun is probably my very favorite of all of the guiding principles of our anthology. Before the ink was dry on the first draft paperback copy of the anthology, we were already making plans for what we wanted to see happen next. The anthology alone was never intended to be the "end" of this journey; it was our beginning.

This is NOT in any way to diminish the significance of the anthology. *All the Weight of Our Dreams* is the first-ever anthology about and by Autistic people of color; quite literally, it made history. It was an important start, a groundbreaking milestone and we remain extremely proud of it. But there is so much more that we want to do to with and for Autistic people of color; the anthology helped us to realize that and to begin dreaming yet again.

It's hard not to feel heartbroken when you mail a contributor's copy to one of the brilliant writers in the anthology—and that copy comes back to you in the mail undelivered because that writer, that amazing, talented Autistic PoC, is now homeless. It's hard not to feel frustrated when you see yet another link to a fundraiser for an Autistic PoC who has been fired from their job again due to racism and ableism and now they are struggling to pay their rent. There are so many stories; there is so much need; there are so many voices still unheard.

We've succeeded in capturing a snapshot of our reality as part of the global community of Autistic PoC. One of its central themes is an illustration of intersectionality. Not in the way it is frequently misunderstood and misused, but as Dr. Kimberle Crenshaw, the Black scholar, attorney, and womanist (among her many honorifics) who is the creator of the term [2], intended: a way of describing the overlapping effect of existing with multiple oppressions/marginalizations—like Autistic PoC do. It's never "just" autism for us, and it's never "just" race. Intersectionality, for us, isn't an intriguing concept to have philosophical debates about; it's our real lives.

And in these real lives, we face unique challenges, but also have unique strengths. It is our hope to be able to impact individuals as well as our community in a meaningful way through this anthology. We have already started to do so, in fact: with a portion of the proceeds we have earned from sales of the anthology as well as an external grant, in summer 2018 on the one-year anniversary of our anthology's publication date, we have launched the first-ever Fund for Community Reparations for Autistic People of Color's Interdependence, Survival, and Empowerment (https://autismandrace.com/autistic-people-of-color-fund/). The fund, which is operated by AWN Network and managed by the anthology editorial leadership team, provides direct financial support to Autistic people of color through individual microgrants of amounts between $100 and $500. We have assisted individuals in purchasing medicine, food, assistive technology, educational workshops, housing costs, legal fees, and more.

We also have plans to develop a few special editions of the anthology in the future, including an audiobook version in as many of our own voices (for speaking contributors) as possible and a version with full-color artwork from the anthology. Additionally, we hope to be able to obtain funding to help support Autistic PoC with scholarships, to help defray the costs of presenting at conferences, and in other endeavors.

The more that I think about it, my involvement with the anthology was no true accident. It was supposed to be this way, and everything is unfolding the way it was designed. Just as I could not foresee becoming part of it, now I can't envision a future where I am not a part of this life-changing project. It isn't just a book; it's truly a dream come true that has the potential to change hearts, change minds, and hopefully continue to change lives.

References

1. Brown, L. X., Ashkenazy, E., & Giwa Onaiwu, M. (Eds.). (2017). *All the weight of our dreams: On living racialized autism.* Lincoln, NE: DragonBee Press.
2. Crenshaw, K. (1989). Demarginalizing the intersection of race and sex: A black feminist critique of antidiscrimination doctrine, feminist theory and antiracist politics. *University of Chicago Legal Forum, 139,* 139–167.

Open Access This chapter is licensed under the terms of the Creative Commons Attribution 4.0 International License (http://creativecommons.org/licenses/by/4.0/), which permits use, sharing, adaptation, distribution and reproduction in any medium or format, as long as you give appropriate credit to the original author(s) and the source, provide a link to the Creative Commons license and indicate if changes were made.

The images or other third party material in this chapter are included in the chapter's Creative Commons license, unless indicated otherwise in a credit line to the material. If material is not included in the chapter's Creative Commons license and your intended use is not permitted by statutory regulation or exceeds the permitted use, you will need to obtain permission directly from the copyright holder.

Part III

Entering the Establishment?

19

Changing Paradigms: The Emergence of the Autism/Neurodiversity Manifesto

Monique Craine

Throughout this chapter, I will be looking at the actions which led to the development of the Labour Party's Autism/Neurodiversity Manifesto in the United Kingdom (U.K.). I will explain where we are currently with the manifesto and what comes next. I will discuss why I chose the approach I took and who was helpful in moving the cause forward and I will also discuss some of the issues and barriers we encountered on our way. I will also define the terms neurodiversity and neurodivergent but firstly I would like to briefly introduce myself.

A Bit About Me

I am a fifty-year-old, married, mother of three; I am also a graduate and am currently in the early stages of launching a new self-employed venture.

M. Craine (✉)
Cwmtwrch, UK
e-mail: monique@mccas.co.uk

Prior to embarking on self-employment (for a second time), I had been employed in a variety of different job roles as well as having time at home raising the children; I had worked in everything from waitressing and sales, to running as a candidate for local elections and getting elected as a Liberal Democrat Borough Councilor, with all the responsibilities that go with that.

That's the positive "sales" version of me. There is another side to me which has coexisted alongside this resume version of me. During this time I have battled low self-esteem, severe anxiety attacks, depression, and the feeling that I just didn't fit in since I was in my early teens. I left school with hardly any qualifications, believing I had failed. I was faced with a career advisor who told me I was officially classed as "less able" and that I would never be able to gain a standard high school qualification; she told me I would most likely not be able to continue with education past age sixteen. I was basically told I was too stupid to ever learn. I was effectively written off at school.

Back then it was just accepted that I was one of those "slow" children. I often refer to my eleven years of compulsory schooling as an eleven-year assault on my self-esteem. I tended to disconnect in class, to allow my thoughts to focus on other more interesting details. The lessons did not interest me, and I struggled with basics like spelling and reading aloud. I knew I was different from the other children, not just in the way I thought but also in the way I moved, my behaviors, and even in the way I perceived the world around me.

From the age of five, I learned to mask (to pretend I was just like everyone else). At school I became selectively mute for a while, as I knew that if I talked people would call me weird; my solution was simple: "don't talk." I knew that my school work was not as good quality as the children who sat beside me so from age five I would hide my work so that the teachers and other pupils didn't notice my mistakes. I would then spend hours rewriting it at home so that I wouldn't be told off. In a way it is no surprise that I went through school without being diagnosed with any conditions. I did after all go out of my way to hide all my difficulties from my teachers.

Late Diagnosis—Diagnosing Difference

Although I was not diagnosed with any conditions as a child, I was finally diagnosed with Dyslexia and scotopic sensitivity (light sensitivity), in my late twenties while studying at university. The knowledge I gained about how my dyslexic brain worked enabled me to stop punishing myself over the things I couldn't do and to instead discover new ways to work that took my different neurology into account.

I found that by adapting the way I worked, I could achieve much better results than I could while hiding my difficulties. I also found that I started to appreciate the way my visual memory worked; I had never realized that there were people who lacked my ability to visualize things. The way my visual memory works may make it an issue when it comes to interpreting two-dimensional letters on a page that represent a word, but it is pretty awesome when you can visualize processes in high definition, slow motion, and zooming into the detail. Once I realized that many dyslexic people have this same kind of high-definition visual memory. I realized that dyslexia gave me something awesome that I would not want to live without. Although I still struggled with reading and writing consistently, the simple truth is, I would rather have the amazing visual memory I have than be able to read and write with ease. Thus I felt proud to be dyslexic; I accepted I would need to put strategies in place to help me with the difficulties, but no way would I want to trade it!

I was very positive about being dyslexic even though I do find it a little embarrassing when my children by age six have found themselves correcting my reading to them at night.

The simple truth is I have never overcome my dyslexia, I have never learned to read accurately or consistently but I did learn to read slowly and repetitively for accuracy. I prefer to use specialist software to read things to me to ensure I am taking the correct meaning from text but given time, I can read. I was very positive about being dyslexic and although I still struggle with reading and writing, I am happy that with the right tech I can perform certain roles just as quickly as someone who can read and write fluently and is not dyslexic.

Dyslexia explained a lot in regards to the difficulties I encountered at school, but it didn't explain all my difficulties. After receiving my dyslexia diagnosis in 1997 (age 27), I still continued to seek medical support for my mental health issues. Anxiety was my "natural" state, and my immune system was so low I would catch everything. I was prescribed medication for the anxiety but it did nothing for me. I still struggled with my coordination—I was falling over every day, I was banging myself and breaking things constantly. I was struggling to maintain relationships that had meant the world to me for much of my life and I was unable to keep a job.

In 2012, I was diagnosed with dyspraxia/Developmental Coordination Disorder (DCD). Finally, I understood where my coordination issues and problems with movement came from. Again, post-diagnosis, I was able to create strategies to help me manage some of my difficulties. As before, I have not overcome being dyspraxic, I still have difficulties daily, but now that I account for my dyspraxia I suffer with fewer injuries.

It was also around this time I became active on the internet. I wanted to learn everything there was to learn about the way I functioned so that I could get my life back on track. Dyspraxia accounted for so much but by now I knew there was more than just dyspraxia and dyslexia going on, but I could still get no guidance on how to proceed. In 2013 one of my family members was put on the autism assessment pathway and once the doctors realized that there were cases of autism within my family I was able to get on the pathway for an Autism Spectrum Condition (ASC) assessment.

Blogging and Other Internet Activities

When I first started blogging in 2013 I was still pretty new to my own diagnoses and I had not yet been diagnosed as autistic. Through my early blogs I repeated what I had been told by professionals. I used their terminology about the conditions but I focused on the positives. I had a "can do" attitude and wrote about ways of improving areas we struggled with.

I believed that my dyslexia and DCD/dyspraxia were just down to the way I was wired. I was always going to have to take my differences into account but more often than not, the root cause of my difficulties were social barriers and my own poor mental health as my anxiety was a constant battle for me. I followed quite a few dyslexia groups on Facebook (FB), and would post supportive comments on parents' posts who were asking for help.

After getting my dyspraxia diagnosis I joined a few FB groups for dyspraxia too. My favorite group was called "Dyspraxia - Dyspraxic adults surviving in a non-dyspraxic world" (https://www.facebook.com/groups/dyspraxiainadulthood). When I joined the group it only had about 400 members and it was very active as the founders had been using the group as research for their book of the same title.

In this Facebook group, I was meeting people with multiple neurodevelopmental diagnoses. Hearing their stories made me realize that my mental health issues, the burnout, the inertia, depression, and the constant anxiety I experienced, could have been brought on by other undiagnosed co-occurring neurological differences, as opposed to being due to a mental health illness. In the end I started my own blog site where I wrote strategy guides so I could post them to the group.

It's worth noting that when I started blogging it was mainly the adult dyspraxic online community I was talking to. However, I soon realized that my tips were not just relevant to dyspraxics, but to a lot of other groups too. The members who had other co-occurring developmental conditions were sharing my blogs in their other groups and similar feedback was coming from them. It was when my blog post on "Cleaning Your Home Made Easy" [1] went viral within the ADHD groups I realized our community had been segregated by a medical model which insisted on separating us from our natural peers and mentors.

The feedback I received for my blog surprised me because it had its biggest success within the ADHD groups. Neurodivergent (ND) adults were sharing it with "before and after" pictures of their houses. It was amazing and to this day I still get comments on how it has changed people's lives. This is the main reason I started talking to and about the Neurodivergent community instead of the segregated individual dyslexia, DCD, ADHD, Autism, etc., online groups. It is why I work toward unifying all the ND minority groups under one more natural banner.

By 2014 my blogs were really successful among the online ND communities in general. I was a well-known figure in the Dyspraxia (DCD) groups but my blogs were not going down as well with the autistic groups as they were with other ND groups. I was constantly getting negative feedback from autistic individuals. In the main, they loved my content but were frustrated by my terminology. Some of these people seemed unnecessarily aggressive over little things like my using person-first language or my using pathologized language to describe neurodiversity. They objected to my saying "person with autism," and for incorrectly using the term *neurodiversity*. A few even trolled me to some extent, posting every time I used any terms incorrectly. Despite these issues, the autistic individuals still seemed to like my content and as the general content was being well received I continued to write my guides and opinion pieces, but if you read my blogs in chronological order you will see how I slowly became aware of the issues surrounding the terminology I had used and over time totally changed the way I wrote about neurodiversity.

Although these autistic people were to some extent trolling me as they would post comments in every group I belonged to, they were not being nasty about it. They just felt the need to correct me every time they felt my terminology ruined an otherwise excellent article. I now consider many of these people among my best friends and am eternally grateful to those who first linked me to the work of Nick Walker. As that was the first time I came across a definition for neurodiversity which I actually understood and agreed with.

Defining Neurodiversity

Walker explained the history of the Neurodiversity movement, he had researched the subject in-depth and concluded that:

> Neurodiversity is the diversity of human brains and minds – the infinite variation in neurocognitive functioning within our species. [2]

Walker also describes "how an individual cannot have a neurodiversity. That if someone's neurology diverges significantly from the typical majority, the correct term to use for them is neurodivergent."

Walker's definition of neurodiversity made sense to me, it showed that autism, dyslexia, Tourette's syndrome, and other similar neurological developmental conditions were naturally forming variations within the human species and those of us with those developmental neurological brain-types were simply neurodivergent. We weren't broken or wrong, just different and thus had very different needs from most.

Understanding the terms *neurodiversity* and *neurodivergent* myself was only half the battle. I still had the problem that I wrote in person-first language, "person with autism," and I still used medical language to describe our differences—in terms of deficits and impairments. It was again the work of Walker who helped me understand why I struggled to convey a positive message regarding my own neurology.

In another article written by Nick Walker they state:

> When it comes to human neurodiversity, the dominant paradigm in the world today is what I refer to as the pathology paradigm. The long-term well-being and empowerment of Autistics and members of other neurological minority groups hinges upon our ability to create a paradigm shift – a shift from the pathology paradigm to the neurodiversity paradigm. Such a shift must happen internally, within the consciousness of individuals, and must also be propagated in the cultures in which we live. [3]

In my early blogs I had followed the pathology paradigm, defining our differences by deficits and impairments. After reading the work of Nick Walker I got it. The pathology paradigm relied on a medical model, a system based on our (observable) presenting behaviors and which used a tick box diagnostic system, depending on whether our issues are mainly related to one cluster of ND-presenting difficulties or another. Having become an active member of quite a few online ND groups it seemed that the "main" diagnosis we got depended on the specialist we were first sent to in order to have our neurological differences diagnosed. I found that most of the group members in all the separate, segregated groups had co-occurring conditions and overlapping qualities. One diagnosis alone could not seem to define anyone's particular developmental history or help them find strategies suitable for their personal profiles.

In the UK all developmental ND conditions are usually diagnosed and assessed by different specialists. Each specialist, be they a psychiatrist, educational psychologist, speech and language therapist, occupational therapist, or other, assesses each condition in isolation. Their knowledge and understanding of neurodiversity varies, and as a result we are not often sent for follow-on assessments for other presenting ND differences.

My mission had been to find out as much as I could about the different types of neurological development but I soon discovered that if you have multiple diagnoses you have to join multiple charities and groups to gain information relevant to you. I had by this point joined three leading charities and at least twenty different Facebook groups. I wanted information relevant to me in regards to my own dyslexia, dyspraxia/DCD, and autism which I had now been put on the assessment path for. It's a maze just trying to find others who have traveled your path but through the internet I was finding them, and they were finding me.

I stopped blogging for quite some time after discovering the work of Nick Walker, as I tried to research more on the topic of neurodiversity and the many different types of developmental neurodivergent conditions but it was also around this time I joined Autistic UK (https://autisticuk.org/), the leading autism organization in the UK that is run by, and for, autistic people. Autistic UK was engaged in the wider ND movement beyond autism. Although its main focus was in autistic rights, it was inclusive of all the other conditions, whereas most organizations would only deal with the one condition mentioned in their name. Autistic UK would look at the whole person and the many different ways they might function to ensure all their human needs were being taken into account. This was the main reason I wanted to join Autistic UK, as I wanted to find one organization who were keeping informed about issues affecting all ND groups.

By 2015 I had finally been diagnosed as being on the autism spectrum myself. I was by now a prominent and active member within many of the segregated ND online groups, sharing my blogs with each group separately. My business was not very successful in monetary terms but it was highly successful for my self-esteem, as I received comments and messages daily from grateful readers. I had just been shortlisted for the National Diversity Awards as a Positive Role Model for people with disabilities and it was at this point that I realized I needed to do more for the communities than just post top tip guides and opinion pieces on Facebook.

From Dyslexia to Autism—Silencing Voices

Talking about my autism diagnosis had different consequences to talking about my other ND differences. Autism was being seen differently to the other neurodevelopmental diagnoses. I had always referred to myself as being "dyslexic" but when I referred to myself as being "autistic" I was told off by parents and professionals as they felt it was insulting to put the condition before the person. Whereas I could talk positively about dyslexia, when I tried to do the same with autism I was being told that I was only a "self"-advocate and was nothing like their child. I was seen as an inspiration to the dyslexic and dyspraxic community but was shunned within the autism community despite being popular among the actually autistic group members.

In my eyes my autism was just a part of my individual brain wiring, no different from my dyslexia or DCD; it just ticked different boxes. There were many carers, parents, and others who worked with autistic people who told me that I had no right to speak on autism because their experience was of working with or raising autistic children who were also nonverbal or learning disabled. I received abuse daily from individuals who felt that I should shut up and not talk about autism positively, as in their view, it was a disease similar to cancer or diabetes which needed to be eliminated.

I soon discovered how hostile the world is toward autistics and realized that our voices are often silenced by people whose opinions seem to hold more weight within society. I also began to realize why the autistic audience had been so hard to please. They were fed up of being talked over.

I made many friends within the adult autistic online community as we all banded together to stop the hate. I had discovered that conditions like ADHD and autism were not seen as favorably as dyslexia, and hardly anyone had even heard of dyspraxia. Despite all these differences being natural forms of neurological development, it seemed as though autistics and ADHDers were being talked over and denied the right to shape their own futures, while dyspraxics remained invisible.

I had a newfound fame within the internet-based ND groups because of my blogs and I had the ability to reach hundreds of people. I knew I needed to do something more.

The Role that Shaped My Actions

In 2002 I had been elected to local government as a Liberal Democrat Councilor and had been able to be part of the process for achieving positive change, albeit at a local level. I saw firsthand how long projects can take from conception to completion when they are done properly. During my time as a councilor, I sat on a number of council committees, and management boards for a number of external organizations. I worked with the police and local communities and set up a number of resident groups. I had gained valuable insight into how to give residents a voice and I wanted to help get these groups a voice.

I couldn't help but wonder whether I would have been elected back then had I known I was autistic and had I disclosed that, but that is a question for another day. The experience I took with me from having been an equal and active member of many forums meant I knew how to speak to the people at the top. Those in political office can really make a difference to our lives, but by 2015 I had no political allegiance of my own so I felt free to contact anyone and everyone that I thought would listen, no matter what party they belonged to.

The Advocates Who Inspired Me

It seemed to me that every day more recently diagnosed dyslexic, dyspraxic, and autistic adults were turning from using the internet to research their own neurological profile, to advocating for others who encountered similar barriers based on their neurology. From there some are unwittingly becoming active frontline campaigners for neurodivergent rights. When I was blogging for the dyslexic and DCD online community I worked with advocates like Sarah Chapman, owner of Operation Diversity (https://

operationdiversityacademy.co.uk/). Sarah champions neurodivergent talents and works hard to create a greater awareness of all developmental neurodivergent differences through her social enterprise.

I also worked with a number of dyspraxic advocates but by far the largest cohort of advocates was from within the autistic community. This seemed to me to be because dyslexic adults were allowed to talk about dyslexia even if they never mentioned any negative sides to it. For example, here in the UK we have the business leader Sir Richard Branson, who speaks openly about his dyslexia and how it helped him achieve greatness in his field. Richard Branson is an inspiration to many dyslexics, but it's harder to find autistic people who can talk about their experience of autism without others who claim to be part of the "autism" community shouting them down or pointing out that the individual speaking cannot or does not speak for their children who are in their views more seriously affected.

There seemed to be a barrier to actually autistic people standing up and talking about the issues of importance to the autistic community. We were not referred to as autism advocates, but as "self"-advocates; denied an opportunity to speak on topics of importance to us; and more often than not, denied payment for delivering talks. There seemed to be two communities at odds with each other where the *autism community* spoke over the voices of the *actually autistic community*. This injustice meant that autistic life coaches, bloggers, and advocates were used to being dismissed and at times attacked simply on the grounds of their autism being different to someone else's, but there were also more serious issues affecting the autistic community.

Through my online work I was introduced to individuals like Emma Dalmayne, the founder of Autistic Inclusive Meets (https://autisticinclusivemeets.com/), who has taken on the battle against quack "cures" from individuals who claim they can help children "recover" from, or be completely "cured" of their autism. These people pray on vulnerable parents who do not understand autism and expose the autistic children and adults to abusive practices, such as drinking bleach in the form of MMS. Emma Dalmayne has led the fight to introduce specific legislation to prevent these types of abusive practices from being sold and used in the UK. Working with individuals like Emma also inspired me to do more than just blog.

Internet-based groups like The Autistic Cooperative founded by Kieran Rose (www.theautisticadvocate.com) have links with autistic campaigners based all around the world. Neurodivergent community members are now finding each other through groups like this, supporting and teaching each other about how to campaign more effectively.

Many of these groups were being formed by autistic campaigners who were also multiply ND, so they were inclusive of members with other related conditions. A network of community leaders was forming on the internet where group leaders were trying to allow neurodivergent voices a safe space to talk. I saw and heard of many injustices through these groups, so I wanted to do something positive. Belonging to all the different ND groups guided my hand regarding how to take my activism out into the real world. I was focused on doing something that would have lasting positive effects for all our people so listening to who group members were talking about was essential to me.

In September of 2015, there had been a shuffle in leadership in the political parties and suddenly many in the ND groups were talking about wanting to reach out to one person. That person was Jeremy Corbyn Member of Parliament (MP), the newly elected Leader of the Labour Party (then and at the time of writing in early 2019, the main opposition party in the UK). In the autumn of 2015 individuals from all the ND groups were talking about how Corbyn seemed different, how if anyone was going to understand our plight it would be him. This intrigued me as it was coming from people who claimed to have lost faith in our political system, from supporters of other political parties as well as from Labour supporters. Jeremy Corbyn seemed to be appealing to many within the different ND communities, because he seemed to be approaching politics differently, so I addressed an open letter to him.

In the open letter I spoke about the emergence of an online-based neurodivergent community and stated that:

> Although we are isolated within our geographical communities we have found others like us and formed huge communities on the internet. We gain support from each other and find ways to overcome hurdles together. [4]

I congratulated Jeremy Corbyn for his appointment of a Shadow Minister for Mental Health and asked:

> Jeremy Corbyn, you have created a Minister for Mental Health and I applaud you for that, it truly is an immensely important post in this era. **Can we ask that you also consider appointing a Minister for Neuro-Diversity to work closely with the Minister for Mental Health?** We are in need of political representation. [4]

An open letter signed by just one person is just a letter, from one person. I wanted to ensure I had the message right and that each of the different ND communities backed my call for the creation for a Shadow Minister for Neurodiversity.

The open letter received hundreds of comments in support in every ND group I shared it in. The ND community then seemed to take ownership of it and started sharing and retweeting it to their own contacts. It wasn't long before the letter had gone viral. I wanted as many comments as possible from members of the different neurodivergent communities before sending it to Jeremy Corbyn MP. It was essential to me that he sees it was the community speaking and not just me.

Between publishing the letter, coming out that I was autistic, my own general incompetence at promoting my paid work, the exploitative nature of the autism world, and my focus having turned to campaigning, I had not been able to make a success of my business.

I felt as though I was being forced to close my business because I could not make enough money, yet I was regularly being asked to attend autism events and speak for free. I remember posting something on my personal Facebook feed about how I was fed up of not having my time valued and how I would no longer support charities and organizations who claimed Autism in their title but who did not value our time enough to pay us.

The Butterfly Effect: Reaching the Right Person

A Facebook friend commented on my post, saying I should meet her friend Janine Booth as she thought we had a lot in common. At this time I was still not very good at introducing myself to new people through the internet; my Facebook account was only used for my real-life family and friends. Yet I was already aware of Janine, having preordered her book "Autism Equality in the Workplace: Removing Barriers and Challenging Discrimination" [5]. I had also seen her website and enjoyed some of her poetry.

Janine was to my mind someone of note in the autistic community, she was a well-known author, activist, and campaigner whereas I was just a new internet blogger. I decided to private message Janine to introduce myself. I attached a copy of the open letter I had published so she knew a little about my own work and aspirations in the field.

Janine got straight back to me and asked that I forward the open letter to the Shadow Chancellor of the Exchequer, John McDonnell MP, as she felt he would be genuinely interested. The Shadow Chancellor is the direct opposition to the Government-appointed Chancellor of the Exchequer, the person responsible for the country's treasury. Although not currently in power having someone of this position on our side seemed essential if we wanted to move forward, I did as Janine suggested.

McDonnell responded to me, genuinely inquiring more about the request I had made in the open letter: the creation of a Shadow Minister for Neurodiversity. From there things moved very quickly. I was invited to meet John McDonnell at Portcullis House, in London to discuss my proposal further but because of my issues traveling I was not able to meet with him immediately. In the meantime Janine Booth had a book launch which John McDonnell was attending and after she asked him if he would support creating an autism manifesto, he announced his support for the creation of a Minister for Neurodiversity and an Autism Manifesto [6].

I was blown away by the announcement and was excited that I would soon be meeting with John McDonnell MP to discuss progressing matters. To be able to attend a meeting in London I needed my daughter to be my travel buddy. I had also asked Janine to come with me to the meeting for two reasons: one was because I was eternally grateful that she had pointed me in the direction of John McDonnell and secondly because I was scared to go in a building where I knew no one (although I had not physically met Janine by now we had corresponded a lot and I felt familiar with her).

Conceptualizing an Inclusive Manifesto—The Meeting

Our communal friend was right—we were certainly both on the same page; our responses to the questions from John McDonnell seemed to be *exactly* the same. During the meeting we discussed many areas of discrimination for the different ND communities we were attempting to represent. John McDonnell asked many questions. What he seemed to want to know was whether a Ministry for Neurodiversity would be enough to achieve the changes all the ND communities needed and whether an autism manifesto alone would be inclusive of all ND groups.

After talking about a variety of different issues in some depth we agreed that in reality much more was needed. We needed to combine my idea with Janine's and work toward achieving the best outcomes for all groups. We discussed how Government policies affect members of the ND community in every aspect of their lives, from health and education to housing, employment, and the judiciary system. By the time the meeting had come to a close, McDonnell had tasked Janine and myself to put together a team of ND individuals to start working on a fully inclusive manifesto to come back and put before him. McDonnell not only supported the "Nothing About Us Without Us" principle we desired, he put it front and center.

The Creation of the Manifesto—Steering Group Actions

Janine and I then put together the Labour Party Autism/Neurodiversity (LPA/ND) Manifesto Steering Group, made up of ND activists, advocates, academics, and campaigners. Our remit was to go to all the ND communities and with the communities' help, design a truly inclusive manifesto to present back to McDonnell.

Over the next few months we worked together to design a manifesto which was felt could bring about the kind of structural changes needed to create a more ND inclusive society. The Steering Group launched the LPA/ND draft manifesto in September 2016, at a fringe event of the Labour Party Annual Conference. McDonnell attended this event and spoke about how he was supporting the creation of the LPA/ND manifesto. He also spoke about how he had first been introduced to the world of autism through his constituency work, and about how he had been challenged by Janine Booth to do something about autism in regards to the workplace at a trade union event he had attended some years ago. My impression was that McDonnell had been thrust into the autism world through his political work long before my open letter arrived in his inbox. You can view John McDonnell's speech online [7].

Austin Harney (LPA/ND steering group secretary), spoke about his own lived experience of autism and why he was fully supportive of the manifesto. He stated:

> When I received the diagnosis, the authorities recommended that I was taken away from mainstream education against my wishes. The top professional medical experts stated that I had sub – normal intelligence that was unfit for mainstream society, education and employment. I was sent to an Autistic compound. [8]

Austin, like so many of us, was able to prove these doctors wrong and is now a civil servant as well as being a member of the national executive committee for its trade union, PCS (Public and Commercial Services) Union.

Janine Booth spoke of the impact Neurodiversity training was having in the workplace, how it was showing real benefits, not just to the ND staff but to other staff and clients too. She then outlined the core principles: the social model of disability, the neurodiversity approach, opposition to austerity, socialism, democracy, solidarity, and the Nothing About Us Without Us principles which had helped us construct the document we were setting out before the delegates. She then outlined the document in full [9].

After the launch, we created a Facebook Page (https://m.facebook.com/LPANDmanifesto/) so that the segregated online communities could add their input easily through their Facebook accounts. Janine ran Neurodiversity training events and we attended meetings and ND events, so that we could between us all obtain more feedback from the many varied, segregated ND communities.

Joseph Redford had been involved in setting up or running many autistic community-building events prior to joining the LPA/ND Manifesto group. He had worked with Autscape, an event run for and by autistic people. Joseph had also worked on Autistic Pride in Hyde Park, an event which started in London and which in 2018 saw autistic people and their families attend Autistic Pride events in cities all over the country. Joseph was able to take the Manifesto and gather valuable feedback from many within the autistic community through these activities and events.

Annie Morris facilitated all our meetings and also set up and monitored the Neurodiversity Manifesto Website for the group (https://neurodiversitymanifesto.com). We also were lucky to have noted academics available to the steering group. Dr. Damian Milton and Dr. Dinah Murray were able to help us research any controversial issues in more depth.

In 2017 the steering group was in the process of gathering feedback to ensure all issues had been considered in the draft document. We wanted to ensure we had not left any groups unrepresented and we wanted to do a good job. We were not due for a General Election in the UK until 2020 and we wanted to produce a document that was well-thought-out so the Labour Party would have no problems adopting it before the next General Election. But this all coincided with massive political change taking place in the UK as the vote to leave Europe had taken place in 2016 and the political landscape was changing. The Conservative Government changed its leadership and Theresa May was appointed the new Prime Minister. In June 2017 a snap General Election was called and we again found ourselves at the ballot boxes.

When the election was called the steering group collated all the information we had gathered. The draft manifesto was then adapted and submitted to McDonnell just before the official Labour Manifesto was itself published. All of us on the LPA/ND Manifesto steering group knew there was no time to have the full manifesto included in time for that election, but as McDonnell had seemed to understand our issues and had attended steering group meetings when he could, we hoped that some of our more general aims would have been brought up.

The official Labour Party Manifesto 2017, "For The Many Not The Few" [10], included some of the core principles requested in the draft A/ND manifesto. The Labour Party Manifesto made it clear that the Labour Party desired an autism-friendly UK, that they wanted to implement the social model of disability, and look to resolve disability issues created by lack of accommodations rather than expecting the change to come from the disabled person. It even mentioned neurodiversity.

In September 2017 we were ready to host the Autism/Neurodiversity event at the Labour Party Conference. A copy of the drafted manifesto was left on all the delegates chairs so they could refer to it throughout. A copy of this can be found online [11].

I Chaired this event, and had the pleasure of introducing McDonnell, who this time spoke of how proud he was of the work Janine and the whole of the steering group had put into creating the Manifesto and again stated his full support for it.

Emma Lewell-Buck MP, the only openly dyspraxic Member of Parliament then spoke about what it is like being a dyspraxic politician. Other politicians and members of the steering committee also spoke about how the Autism/Neurodiversity Manifesto would impact the different ND groups and how that could benefit society.

We had talks by dyslexics, autistics, dyspraxics, and a very vivid and powerful talk by Terry Laverty, a core member of the Steering Group, who spoke on the topic of ADHD. Terry spoke about what it is like having a mind that works so differently to most and the discrimination faced by ADHD adults.

The event was very well attended and support for the manifesto was unanimous. One of the comments from the delegates which stuck with me was that they had never been in a room with so many neurodivergent people in one place. It really was an amazing feeling to be in a room full of our actual peers.

Where We Are Now—The Present

In 2018 we ran another successful conference where we introduced our critical appendix on Applied Behavioral Analysis (ABA) which was also incredibly well received [12]. Shortly later Janine and myself reported back to John McDonnell MP, to present the final manifesto to him in detail [13]. John McDonnell fully supports the manifesto and is keen to see the whole document officially adopted by the Labour Party and has explained how we can now go through the right channels to get it approved by the National Executive.

Launch of Neurodivergent Labour

I learned through my time as a borough councilor that all projects take time: there is no guarantee that just because you start a project, you will still be there when it is approaching completion. Therefore, the best projects are the ones you can walk away from knowing that you have already got the ball rolling in the right direction.

I feel proud to know that if I walked away soon I would have left behind me a democratically formed organization who have pledged to promote the A/ND manifesto and to fight for the rights of neurodivergent people. I have faith in this organization being a force for positive change.

On February 9, 2019, I saw politics on the brink of a paradigm shift as it opened the door to neurodivergent voices. John McDonnell again spoke of his support for the work we had done and the ambitions we had to make the Party more accessible to us. As the Disability News Service later reported, John McDonnell said:

> As I keep repeating, this is not about electing a group of MPs who will go off and do it for us. That will never work, it never has and it never will. This is about when we go into government, we all go into government, so we draft our manifesto, we secure commitments from the bulk of the Labour Party through the normal policy making process and then when we go into government, we all work on the detail of the implementation itself. [14]

Hearing McDonnell speak filled me with confidence. Knowing that it is no longer just individual voices trying to be heard, but a movement of people who are uniting under the Neurodivergent banner.

I am eternally grateful that my open letter to Jeremy Corbyn, MP, fell into the hands of Janine Booth. I'm so glad she had the political awareness to make it much bigger than it was, by getting it to the best person to read it, and for pushing for a manifesto. I am eternally grateful to John McDonnell for understanding the need to create a ND inclusive manifesto so we can all be supported, and for allowing us, the ND individuals, to lead throughout the process of inception, construction, and delivery.

I hope that other ND advocates are writing to people they think might help. Despite the internet trolls we have to contend with, there are actually a lot of good people out there and there are some amazing allies still to be made. It is amazing what a simple letter can lead to. My aim was to start a ball rolling, to start something that I could eventually walk away from knowing that the momentum alone would keep it going. I wanted to start a conversation that included us at all levels. What I hadn't realized at the time, was that hundreds of ND advocates are out there doing the exact same thing. One person at a time, perspectives are changing, and paradigms are shifting. As more and more advocates find each other and work together our voices are getting heard.

References

1. Craine, M. (2014, July 5). *Cleaning your home—Made easy* (Web log post). Retrieved from http://needtosay.weebly.com/blog/cleaning-your-home-made-easy.
2. Walker, N. (2014, September 27). *Neurodiversity: Some basic terms & definitions* (Web log post). Retrieved from http://neurocosmopolitanism.com/neurodiversity-some-basic-terms-definitions.
3. Walker, N. (2013, August 16). *Throw away the master's tools: Liberating ourselves from the pathology paradigm* (Web log post). Retrieved from http://neurocosmopolitanism.com/throw-away-the-masters-tools-liberating-ourselves-from-the-pathology-paradigm.

4. Craine, M. (2015, September 16). *An open letter to Jeremy Corbyn* (Web log post). Retrieved from http://needtosay.weebly.com/blog/an-open-letter-to-jeremy-corbyn.
5. Booth, J. (2016). *Autism equality in the workplace: Removing barriers and challenging discrimination.* Philadelphia, PA: Jessica Kingsley Publishers.
6. Weaver, M. (2016, May 31). Labour to appoint a shadow minister for neurodiversity. *The Guardian.* Retrieved from https://www.theguardian.com.
7. McDonnell, J. (2016, September). *LPA/ND launch speech* [Video file]. Retrieved from https://videopress.com/v/5C8MnEvD.
8. Harney, A. (2017, June 3). *A post from Austin Harney* (Web log post). Retrieved from https://neurodiversitymanifesto.com/2017/06/06/a-post-from-austin-harney-a-member-of-this-steering-group.
9. Neurodiversity Manifesto. (2016, September 22) *Neurodiversity manifesto: Labour Party launch* (Manifesto). Retrieved from https://neurodiversitymanifesto.com/2016/09/22/neurodiversity-manifesto-labour-party-launch.
10. Labour Party. (2017). *For the many not the few* (Manifesto). Retrieved from https://labour.org.uk/wp-content/uploads/2017/10/labour-manifesto-2017.pdf.
11. Labour Party Autism/Neurodiversity Manifesto Steering Group. (2017, September 15). *The Labour Party Autism/Neurodiversity Manifesto (V.2, 2017)* (Manifesto). Retrieved from https://neurodiversitymanifesto.com/2017/09/15/the-labour-party-autism-neurodiversity-manifesto-v-2-2017.
12. Milton, D. (2018, September 15). *A critique of the use of Applied Behavioural Analysis (ABA): On behalf of the Labour Party Autism/Neurodiversity Manifesto Steering Group* (Manifesto appendix). Retrieved from https://neurodiversitymanifesto.com/2018/09/15/labour-party-autism-neurodiversity-manifesto-2018-please-see-our-final-draft-on-page-3.
13. Labour Party Autism/Neurodiversity Manifesto Steering Group. (2018, September 18). *Labour Party Autism/Neurodiversity Manifesto: Final draft version* (Manifesto). Retrieved from https://neurodiversitymanifesto.com/2018/09/18/labour-party-autism-neurodiversity-manifesto-final-draft-version-2018.
14. Pring, J. (2019, February 14). Launch of Neurodivergent Labour 'could be milestone in fight for rights and equality'. *Disability News Service.* Retrieved from https://www.disabilitynewsservice.com.

Open Access This chapter is licensed under the terms of the Creative Commons Attribution 4.0 International License (http://creativecommons.org/licenses/by/4.0/), which permits use, sharing, adaptation, distribution and reproduction in any medium or format, as long as you give appropriate credit to the original author(s) and the source, provide a link to the Creative Commons license and indicate if changes were made.

The images or other third party material in this chapter are included in the chapter's Creative Commons license, unless indicated otherwise in a credit line to the material. If material is not included in the chapter's Creative Commons license and your intended use is not permitted by statutory regulation or exceeds the permitted use, you will need to obtain permission directly from the copyright holder.

20

From Protest to Taskforce

Dinah Murray

The National Autistic Taskforce's (NAT) main focus is on autonomy and justice for people who don't use speech to meet their needs; that was also the main focus of the APANA story, and that is where the next stage of this narrative starts, as one of its key players is Virginia Bovell who was on the parliamentary advisory group that learnt about neuroleptics from Wen Lawson and myself. This section is about how the NAT came to be and what lessons can be learnt.

Autistic Influence on the Emergence of Autistica

At that All-Party Parliamentary Group on Autism (APPGA) meeting mentioned above, and after it, my concession that in some cases at low doses

---

D. Murray (✉)
London, UK

some people find risperidone helpful was clearly *a vital step towards credibility* (though not liked by the APANA parents at the time). Virginia, founder of an applied behavioral analysis (ABA) school and achiever of British version nicey-nice ABA, was also concerned about eugenics, so a while later we met again to talk about that and I met her son Danny. Some years passed, then one day in 2008 she and I did a double take in the street and she asked me in for a cuppa (tea). We had a very pleasant chat and exchanged phone numbers. Despite our very different views of behaviourism, we had much common ground.

Just a week later I got a very disturbed phone call from Virginia. A new CEO of her school (Treehouse), unaware of their bad name had invited Bob Wright, founder of Autism Speaks, to launch a new series of annual lectures on autism. To many people it was already clear how distorted their deeply medical model of autism was, with its analogy to cancer as a problem to be wiped out as soon as possible, including by genetic intervention. So Treehouse needed help and we built on our common ground despite a difference of view that could have ruled that out.

The solution I proposed was to: establish a creative autistic presence; give autistic people a chance to say what people most need to hear about autism; have the widest possible exposure; rebalance the specific event to diminish Autism Speaks' impact. This created alliances and generated obligations. Treehouse gave me funding for a professional editor, for two-minute videos from the Autistic Self Advocacy Network (ASAN), members of my Posautive Youtube group, and Autscape (see Buckle, Chapter 8) members as well as from personal contacts. Two thousand *Something About Us* DVDs (*Something About Us* [1] were made and distributed—free and copyright-free (now part of the exhibition at RightfulLives, www.rightfullives.net/Community-of-Perspectives.html). At the event in London's City Hall, everyone including the visiting Wrights was asked to flap not clap. An autistic woman, Anya Ustaswewski, cogently responded, and an audience with many invited autistic people (see the videos at RightfulLives). There was a small demo outside with these signs (see Figs. 20.1 and 20.2).

Another result was that people were indebted to the autistic input and Virginia introduced me to key people who recognized the debt (I was not paid myself). Those included Hilary Gilfoy of whom more below.

Fig. 20.1 Anti-Autism Speaks Logo designed by Dinah Murray

Fig. 20.2 Another Anti-Autism Speaks Logo circulating in 2008, anonymous

This event strengthened my standing as a reasonable person (not all may agree) again because I had conceded some ground myself, i.e. that the UK behaviorists were not themselves devil's spawn. The whole ghastly visit—perhaps with a bit of help from our well-found critique—contributed to a painful and difficult process as the UK organization that had been paired with the US Autism Speaks severed the partnership and reinvented itself as the strictly British Autistica (which now has a Director of Science who is autistic himself). I kept up good relations with Hilary Gilfoy in person over the next years as she seemed to me a fair-minded person who was going to some lengths to distance and separate her organization—funded by Dame Stephanie (Steve) Shirley in the UK—from the Americans.

Autistic Input into the National Autism Project

A few more years passed and in 2014, Dame Steve, who had used her unexpected great wealth (see her autobiography, *Let It Go* [2]) to give very many millions of pounds toward fathoming what autism is all about, decided to commission the "National Autism Project" (NAP) to study whether anything useful had emerged from all these decades of autism practice and research, much of it funded by her.

Thus, in 2015 Ian Ragan and Elizabeth Vallance (from the original Autism Speaks UK) approached me to join the NAP Strategy Board. I felt honored, but it mattered to me that if I said yes I wouldn't just be there as "window dressing." As I explained, my experience as a trophy autistic person at the National Autistic Society had been recent and bruising. I also did some agonizing over the generous benefactor's past involvement with Autism Speaks and her own attitude to autism; but she had funded the first online autism conference, in 1999, to which I'd contributed with Mike Lesser and the Webautism course which I helped develop and taught on at Birmingham University—and Autistica was being transformed. Also, they agreed to everything I asked for.

I asked for: communication support for attending Board meetings (in the shape of Damian Milton), as I knew my ability to speak up in a timely manner was limited at best; I was also concerned that I needed my own 'sounding board' and input from beyond my own limited perspective. Happily both these ideas were accepted and I constructed a strong advisory panel with a wide range of both academic and practical knowhow, and I was able to come to Strategy Board meetings with the unfailingly articulate (and like-minded) Damian. Hilary Gilfoy, who had guided Autistica's disentanglement from Autism Speaks, was not technically on the Board, but took excellent minutes at Board meetings and was perceived as a supportive presence by both Damian and myself (and, I think, by her old friend Dame Steve) owing to her calm friendliness.

Having someone as steady as Damian by my side, who also had a fantastic grasp of all the key issues, transformed my capacity to be of some use at the meetings. Eventually everyone was treating us both as equal Board members, and he was contributing freely on the spot in a way I cannot perform myself. There were few occasions when our views diverged, which obviously helped—and at the final Board meeting Damian couldn't attend and I managed to contribute quite fluently myself as I had learnt to trust the people there. How did that happen?

It didn't start too well. A lot of work had been done and decisions taken before the first Board meeting took place—that included drawing up a list of Experts, all of them professionals from academics to psychiatrists to charity bods. I queried why the Autistic Advisory Panel (AAP), who I knew to be deeply knowledgeable in the field, did not also count as "experts." I went and met and talked to the report researchers Professor Knapp and postgraduate Valentina Iemmi quite early on, and I think opened their eyes to how much disability can be created by a hostile environment and what that might mean vis-à-vis autism. They appeared genuinely interested. Even so, by the summer of 2015 I was beginning to think of resigning because of the way the AAP members were lumped together as "Dinah's panel" and their individual great expertise disregarded. However, Damian, AAP member Catriona Stewart, and I had a chat at Autscape and Catriona argued for the Panel to have a face-to-face meeting, which we later did, with the NAP Project Leader, the NAP Chair, and Hilary in attendance. That was the beginning of the real listening.

Gradually all the people on the Panel became distinct and valued contributors to the NAP. The interesting and open-minded other members of the Strategy Board began to hear favorable things about us. Two of the AAP members were turned from pawns to queens and added to the Experts list—Drs Yo Dunn and Catriona Stewart. Cat was their Scottish specialist and Yo was far the most effective and knowledgeable expert they had when it came to calling the Government out on its own laws. In effect she became the NAP's warhead, wheeled into many discussions with senior civil servants to blow them away with her detailed and accurate legal knowledge.

By the time the Project report, *The Autism Dividend*, was launched in early 2017, the large contribution of members of the Panel as well as our "experts" was being explicitly recognized. The report was repeatedly revised and improved by our critical input—and saved from some serious failings—which resulted in Damian and myself being honored at the report launch in a House of Lords venue, as "Productive Irritants" by its lead author, Martin Knapp. Some fundamental differences from his earlier report on "the cost of autism" were that the burden/disease concept had been replaced by explicit recognition of autistic potential and of the varied barriers that prevent it from being realized; it also highlighted a very poor evidence base for most practice, resulting from widespread very low research standards (see autistic researcher Michelle Dawson on Twitter, @autismcrisis, for much more about those).

At that year's Autscape (see Buckle, Chapter 8), five members of the AAP were there, and the National Autism Project's willingness to listen and take us seriously was feted and rejoiced in: the final report is much admired. We are all proud to have contributed to its excellence. Widely seen as the highlight of the 2017 Autscape, Yo Dunn gave a stunning, passionate talk about what she called "The Other Half" and what is more fluent autistic people can and should do about the vast numbers of autistic people who are not articulate but depend on frequent or full-time support. Privately we discussed the idea that if some sort of future for the AAP was going to emerge, its point could be to focus on the Other Half.

In parallel with these discussions, unknown to us, Dame Steve was having some thoughts of her own. Those led her to entrust me with the generous sum of £100,000 to fund a continuation and transformation of the NAP's AAP. She was clearly pleased we chose to focus on people, like her late son, Giles, who need the most communication support. The new body was named the NAT and acquired the strapline: Bolder Voices—Better Practice.

The National Autistic Taskforce

With Damian Milton as project leader, we conceived the Taskforce's main practical aim to be to turn more of the promises encoded in weak and toothless legislation into realities, for people who are poorly placed even to recognize let alone defend their own rights and interests and whose parents are often cut out of discussions when their offspring are teenagers.

We are identifying specific targets and crucial steps along the way, as well as potential obstacles and ways to get around them. We have small focused working groups (GNATs, or Groups of the NAT) on changing practice in the target areas. We are also building up a wider network of autistic individuals and groups across the British Isles so we can draw on a wide range of expertise with as broad and well-informed views as possible. Thanks to this terrific bunch of committed and knowledgeable people, we have been already able to interact constructively with agents of government and other stakeholders, and thanks to Yo Dunn in particular, the Taskforce has drafted several well-supported and closely argued responses to relevant government consultations on behalf of NAT (see nationalautistictaskforce.org) and created the NAT Independent Guide [3].

We think we are well-placed to be seen as carrying some weight, with an executive made up of Kabie Brook, Leneh Buckle, Yo Dunn, and Damian Milton as project leader; and the NAT's history and status mean we have some enviable introductions and contacts. We think our connection with the National Autism Project means that, from an establishment position, we are seen from the start as both credible and significant. Big charities that say they are speaking for autistic people always have their own survival as top priority.

Lessons Learned

What lessons can we learn from this narrative, stretching from the mid-90s to the present day? The process involved creating, discovering, and using community of interest—i.e. shared values and passions—among a diverse range of people whose interests and views outside our specific common purposes often differed widely. Working alongside other autistic people may take quite a lot of additional work (by all concerned) to ensure that communication is effective and perceptions of all sorts are factored in, allowed for and not seen as insuperable barriers, even if there's some potentially painful cognitive dissonance. That said, having autistic comrades along when entering any lions' dens can in my experience make all the difference between being able or not able to communicate effectively. Our impact partly depends on pragmatic adaptation, but our strongest suit is being people with the passion and commitment and indifference to hierarchy to persevere obstinately against the odds. Be an opportunity hound, don't miss a chance!

I have always made a point of assuming everyone has a good heart, as Kabie Brook of Autism Rights Group Highlands (ARGH) recently pointed out to me in relation to the NAT—only once or twice has this precept let me down. In every campaign I found that goodwill, fellow feeling, and knowhow were freely shared, and personal connections were crucial. *Expecting and also returning* tolerance, understanding, and acceptance from others has worked well to further aims. I found it's best to expect mistakes, delays, miscommunications, so as not be "thrown" by these: they are inevitable and it is usually pointless or harmful to attach blame (of self or others) to any of them. If at all possible, stay polite, work on the common ground and remember that though changing practices and attitudes can take decades, lots of small changes really can add up to big ones.

On a more cynical note, remember that when reputations are at stake the power imbalance shifts if those with least power have a voice that can be heard. That's why digital inclusion is so vital and a strategic goal of the NAT. On we go.

References

1. Murray, D., & Benstock, J. (2008). *Something about us* [Documentary].
2. Shirley, S., & Askwith, R. (2012). *Let it go: The memoirs of Dame Stephanie Shirley*. UK: Andrews UK Limited.
3. National Autistic Taskforce (NAT). (2019). *An independent guide to quality care for autistic people*. https://nationalautistictaskforce.org.uk/wp-content/uploads/RC791_NAT_Guide_to_Quality_Online.pdf.

Open Access This chapter is licensed under the terms of the Creative Commons Attribution 4.0 International License (http://creativecommons.org/licenses/by/4.0/), which permits use, sharing, adaptation, distribution and reproduction in any medium or format, as long as you give appropriate credit to the original author(s) and the source, provide a link to the Creative Commons license and indicate if changes were made.

The images or other third party material in this chapter are included in the chapter's Creative Commons license, unless indicated otherwise in a credit line to the material. If material is not included in the chapter's Creative Commons license and your intended use is not permitted by statutory regulation or exceeds the permitted use, you will need to obtain permission directly from the copyright holder.

21

Critiques of the Neurodiversity Movement

Ginny Russell

Preamble

I am going to recount some of the main reproaches to the movement, as I understand them, and show how these are sometimes answered or addressed by our contributors. This does not claim to be a "comprehensive" account of critiques, as my knowledge of the movement is incomplete, and like others' understandings, my writing is situated by and limited by my own reading and experiences [1]. Nevertheless, I believe it is important to be aware of critiques, to engage with criticism, and openly debate, defend, or modify one's position, in both the political and academic spheres.

The start of my chapter concerns critiques that apply to identity politics more broadly: that they dichotomize allied groups into factions (this prevents smaller identity groups from linking up, causing rivalries and discord). Sociologist Charles Derber asserts that identity politics does not include a broad critique of the political economy of capitalism, instead

G. Russell (✉)
College of Medicine and Health/College of Social Science
and International Studies, University of Exeter, Exeter, UK
e-mail: g.russell@exeter.ac.uk

© The Author(s) 2020
S. K. Kapp (ed.), *Autistic Community and the Neurodiversity Movement*,
https://doi.org/10.1007/978-981-13-8437-0_21

focusing on reforms. In a version of "divide and rule" argument he suggests that fragmented and isolated identity movements have allowed for a far-right resurgence [2]. There is also friction within identity politics over definitions of who is included as "in" a particular group. The main critiques of the neurodiversity movement are then listed as follows: first, the movement has been accused of being unrepresentative of all people who are "neurodivergent," and specifically unrepresentative of more impaired people on the autism spectrum (a criticism made by some clinicians, autistic people, and parents). Second, it is said arguments made by the movement are reductionist, promoting a genetic/brain-based understanding of autism (a critique made by academics in social sciences, history, and philosophy of biology). This may deflect attribution of personal responsibility for behavior to the brain. Third, there is a criticism along the lines of the well-known declaration that "the master's tools will never dismantle the master's house" [3].

How Identity Politics Dichotomizes Would-Be Allies

Throughout this collection, contributors use the terms "neurotypical" and "neurodivergent" to denote two distinct groups. Accounts in this collection by Garcia, Neumeier and Brown, and Arnold, amongst others, utilize both terms. Disability scholar Runswick-Cole [4] has pointed out these terms can be divisive, fostering an "us" and "them" mentality. She denounces a dichotomized view of the world where you are either "in" or you are "out".

As we know from many studies of autism and other neurodevelopmental conditions, autism and other diagnostic classes are psychiatric constructs that denote a spectrum: a series of interrelated multidimensional traits. These extend into the subclinical population, therefore many people who do not have an autism diagnosis have autistic traits. This is known by researchers as the "broad autism phenotype" [5–7]. It means there is no clear bimodal distribution separating people with and without autism, so in reality there are not two distinct populations, one "neurotypical" and one "neurodivergent." Instead autistic traits are distributed normally in the whole human population [8], as are ADHD traits [9].

In addition, the tendency to dichotomize runs contradictory to the inclusive definitions of neurodiversity given in the book. DaVanport (Chapter 11) talks about neurodiversity thus: "Neurodiversity soon became something that I intimately understood as the all-inclusive acceptance of every neurological difference without exception. I further came to appreciate that neurodiversity didn't leave anyone out." Greenburg (Chapter 12) defines neurodiversity: "Neurodiversity in my world, is the unquestioned right for all, whatever their neurological makeup, to express what they need or want." She writes: "While I can't define the totality of Neurodiversity even for myself, much less anyone else, I know spaces where I can see it." At another point in her text, Greenburg puts the emphasis on difference, describing "non-autistics whose brains and experiences differ from ours so much" (Chapter 12).

"Neurotypical" (NT) is never used pejoratively in this collection. But in autism online commentaries and debates it has been suggested such critical language is justified by the poor treatment meted out to the autistic community (as well documented here). Tisoncik discusses "autistic superiority" (Chapter 5) which is the tendency to claim autistic people are in some way superior to NT people. She writes, "I am not upset about claims of autistic superiority," explaining that although this is not her position, "we don't need to worry about oppressed groups with little to no power singing their own exclusive praises." An example of such a claim comes from description of NTs from the forum *Quora*: "I'm being cruel I know. It's okay though they have all the power and they have discriminated against me all my life. I'm allowed to make fun of their innate stupidity…they expect everyone to know it because they are the center of the universe. Their hypocrisy on this is near universal" [10].

The issue with this is common to all identity politics and best explained by alluding to another movement—the feminist movement. The argument made is that it is fine to be pejorative about, or insulting to, people that profit from the status quo. But this is the same as a feminist saying pejorative things about men e.g. "all men are <insert pejorative/offensive word here>": a blanket dismissal of all men. Clearly, there are some men who are feminists and sympathetic to the aims of the feminist movement. Plus, many men have characteristics that mean they too are discriminated against: gay men, disabled men, migrants, and so on. To dismiss them all in a blanket way will have the effect of alienating them from the cause. To effect societal change, solidarity between such groups is vital, and those on the left seeking to rebalance the status quo need to support each other's struggles. In addition, "all men are <insert pejorative term here>" is itself a sexist statement. It is discriminatory to make offensive statements about the "out" group, where the struggle should be *against* discrimination and prejudice.

The true target for righteous anger, Runswick-Cole [4] and others have argued, should be discriminatory and disabling societal structures, norms, and practices. It is an issue many activists writing here seem well aware of. Seidel, for example, writes: "I felt that needless discord, demagoguery, and polarization could only exacerbate tensions and undermine advocacy efforts" (Chapter 7).

Standpoint theorists support the idea that people with lived experience have expertise in their own area [11]. This is the theoretical epistemic stance that underpins this volume. It assumes standpoints are relative and cannot be evaluated by any absolute criteria, but makes the assumption that the oppressed (autistic people) are less biased (or more impartial) than the privileged (NT people). This idea that people with lived experience should be given more authority to speak, and make decisions about their own futures, i.e. their voices should be given more weight than others, has been criticized. To illustrate, take the example of female genital mutilation (FMG), a practice widely abhorred. The main promoters of such practices are often the grandmothers of the girls involved who themselves have been subject to FGM. According to standpoint epistemology, such voices (calling for FGM) should trump those of Western medical experts.

Another critique of the NT/ND divide is that "Neurotypical" is a very dubious construct, and by default then so is "Neurodivergent." Is there anyone who is really, truly neurotypical? As Armstrong writes:

> There is no such standard for the human brain. Search as you might, there is no brain that has been pickled in a jar in the basement of the Smithsonian Museum or the National Institute of Health or elsewhere in the world that represents the standard to which all other human brains must be compared. Given that this is the case, how do we decide whether any individual human brain or mind is abnormal or normal? [12]

Of course, you do have to draw a line in the sand when you are demarking yourself into a politically mobilized group. That is, for the neurodiversity movement to exist there has to be a banner "Neurodivergent" under which people can rally. That is required by any group—that you can argue for rights for some people and not others, make the case for services for some people and not others. I will drop the quote marks around "Neurodivergent" for the remainder of this piece, as it is impossible to define an identity-based movement without having a group identity. In order to promote a positive self-identity too, you first need a group-based identity. I understand that. But it is helpful to be aware, in tandem with this, that dichotomizing can also cause difficulties.

Defining Who Is "In": Who Is Neurodivergent?

Another issue that applies to the whole of identity politics is the problem of definition of people in the category. It is not always clear from accounts in the book who is neurodivergent, and who is not. However, this is a crucial issue. If you are advocating for legal protection against discrimination, or arguing for support and accommodations, it is really important to be able to define who that group actually are. If you are not clear who exactly you are fighting for, those rights cannot be operationalized in law.

The nature of brain differences between autistic and non-autistic people is not well-established or well-replicated [13], and many neuroscientific studies of ADHD, Tourette, autism, and other neurodevelopmental conditions have mixed results that are not well-replicated. The reality is that most of these conditions are diagnosed via observation, cognitive testing, or self-report, and not via neurological anatomy or physiology. Not many diagnoses involve brain scans, so the neurological differences of neurodivergent people are not seen but inferred.

Perhaps who is "in" as neurodivergent (ND) could be decided by a medical diagnosis of a neurodevelopmental disorder, or a mental health condition that is thought to be underpinned by neurophysiology, like autism or depression? Neumeier and Brown seem to suggest this when they write that "neurodivergence [includes] psychiatric disability or mental illness" (Chapter 14). But the neurodiversity movement rallies against the "medical model" as Arnold describes it, explaining his journal's aim to "embody a questioning and examining of the prevailing paradigms of autism research" (Chapter 15). Then it is perhaps inappropriate to have group inclusion criteria defined by medical diagnosis. Such a definition would return the epistemic authority to define the group to medics, thus rescinding the notion of "Nothing About Us Without Us" quoted extensively by many contributors. Moreover, some people considered ND may not yet have been diagnosed.

Perhaps, then, self-identification as ND is less problematic? Giwa Onaiwu opted for this approach, "to accept the validity of people's self-identification as stated," when compiling her intersectional anthology (see Chapter 18). Self-definition certainly avoids the problems listed above, but has caused huge divisions in other areas of identity politics. Arguments between radicals in the feminist movement and the transgender activists again provide an example. Some radical feminists have argued that being a woman should be defined by biological sex and being bought up female from birth, whereas the transgender activists have argued that anyone who self-identifies as female is female. The trans-excluding radical feminists assert that decades of rights they have fought for to have women-only safe spaces are now being undermined, if (wo)men who self-identify as women are now allowed into them. This issue was satirized in the UK by a Labour party member who previously made it onto the list of candidates for women's officer because he "identifies as a woman on Wednesdays," under Labour's "self-id" rules [14]. The man has now been suspended from the party.

Self-definition may also exclude some people, who may be part of the "in" group, but may not be aware of being in the group at all. For example, ND people with late-stage dementia, or profound intellectual disability. Others who may be ND don't want to be labeled as such. Activists writing in the collection seem aware of the potential complications arising from dividing people into well-defined groups, and several try to address the problem. For example, Buckle explains the decision for no inclusion criteria whatsoever at Autscape thus:

Having inclusion criteria, e.g. "autistics only" creates suspicion about whether those in the group are really "us" or may be "them," whether deliberately (infiltrators) or by mistake (falsely identifying as autistic). Exclusivity also lends itself to the spread of prejudice and misinformation about the excluded group. (Chapter 8)

Even having read this collection and others, I still am unclear about who exactly is "in," how widely the neurodiversity movement casts its net in defining ND. Does it include just people with autism and other neurodevelopmental conditions like "people with autism, dyspraxia, ADHD," as Garcia states (Chapter 17), or does it further include people with depression, schizophrenia, Tourette's, psychopathology, as Neumeier and Brown suggest in Chapter 14? In this case neurodiversity should be inclusive of neurodegenerative conditions like dementia, and Parkinson's too. The problem is the boundary around who is "in" the ND class and who is "out" is currently not transparent or well-defined.

Representativeness

Several vocal autistic people and parents have complained that the movement is made up mostly of less impaired individuals who do not represent people with more severe problems [15]. I have also heard people comment at autism conferences that persons in the movement are not representative of most ND adults or children, and are not well-appointed to speak for them. My understanding is that the argument is, broadly, parents of more severely disabled children are keen for treatments to ease their children's condition, whereas the neurodiversity movement is seen as anti-cure [13]. Activists counter that the movement does advocate for supports that mitigate weaknesses associated with autism, arguably focusing more on improving access to reliable communication and certainly more on essential services (which mostly go to those with the highest needs) than most organizations and individuals interested in curing autism.

Those parents supporting more medically-oriented models identify the distress and difficulties associated with neurodevelopmental conditions as impairments [16]. Such difficulties lead to problems in functioning and lower quality of life. From this point of view, the conditions that are encompassed by neurodiversity are medical conditions that can and should be cured if possible. Autistic people have also spoken in favor of this more pathologized view: "Many of us aren't high-functioning enough to benefit from depathologizing autism…I still feel autism keeping me from achieving my potential" [17].

It is important to note, however, that many parents are in the movement, including some writing here. The movement allows space for parents and other allies. Sometimes their children, as in the case of Des Roches Rosa, don't have the language skills to engage in conventional activism. So parents are another way in the movement that those who cannot directly represent themselves in formal activism are represented.

Academics have made similar critiques. Ortega [18] argues that so far the movement has been dominated by people diagnosed with Asperger Syndrome and other forms of "high-functioning autism." I have heard a pediatrician-academic dismiss the movement using the same criticism. Casanova, a prominent neurologist writes: "the records that we have at present on neurodiversity are the records of an elite, those that stand at the top…misrepresentation of opinions of the pro-neurodiversity elite as being representative of those at the bottom" [19]. This forms part of a vitriolic attack that claims, "the only thing they have accomplished is the creation of a split in the autism community that allows for themselves and nobody else…Neurodiversity is a social club where many of its participants are non-autistic individuals claiming to be autistics." In another article (also condemned by members of the movement) Jaarsma and Welin make the case that that the neurodiversity doctrine is sensible if it is only applicable to this narrow group "only a narrow conception of neurodiversity, referring exclusively to high-functioning autists, is reasonable" [20].

Activist blogger Hiari, who herself has been given a diagnosis of autism, writes for the critical psychiatry site *Mad in America*. She issues another stinging critique of the neurodiversity movement [21], writing that the movement amounts to no more than:

> A public relations campaign that emphasizes the many positive qualities associated with some presentations of autism—creativity, increased tolerance for repetition, enhanced empathy, superior ability to master content

in specific subject areas, and exceptional memory—while erasing or minimizing the experiences of autistics who are severely disabled.

People at the vanguard of the neurodiversity movement have often been autistic, as this volume testifies, and they have mobilized through making contact online or on email lists, *InLv*, and *autistics.org* being prime examples. The main method of mobilization of the autistic/neurodivergent community therefore inadvertently excludes people who are unable to make contact with each other on computers: "Many on the spectrum can't speak or use a computer" Mitchell asserts [15]. In the current volume, Buckle discusses the efforts that have been made to be as inclusive as possible at Autscape: "we have tried to include some activities that are more accessible to people who don't handle words as well as most of us, with mixed success." She also touches on the practical barriers to achieving full inclusivity and provides a fantastically practical guide to designing autistic-friendly events.

Parents point out that as young children and those with profound intellectual disability cannot advocate for themselves, they as parents must. Activists in the movement counter that parents may have a different agenda to those with lived experience. Activists have stated the case for first-person representation: that elsewhere in society we accept the idea that anyone who speaks for a group should be a member of the group, and by that reasoning any spokesperson for autistic people should be autistic. There are some prominent advocates who themselves or whose child have been diagnosed with intellectual disability or who are non-speaking, some of whom have contributed to this collection. Baggs, who is non-speaking, writes "I'd far rather have Michelle Dawson, Cal Montgomery, Laura Tisoncik, Joelle Smith, or Larry Arnold, speaking on my behalf than these parents" (Chapter 6).

The implication of several writers in the collection is that they are related to other autistic people. Garcia writes of "my autistic brethren" (Chapter 17), and daVanport, "we are linked by a familiar neurology" (Chapter 11). Such statements create a kinship argument: that all autistic people are somehow of one tribe and genetically related, or neurologically similar, so are best qualified to talk about issues that affect the autism community.

The Science and Technologies scholar Silverman critiques the kinship argument in her book *Understanding Autism* [13], explaining that autistic advocates' predicate their claim to represent the autism community in "presumed neurological and genetic likeness to other autistic people" (p. 142). Silverman argues such presumed relatedness lends ethical legitimacy and entitlement to be representatives of the whole group, but points out a contradiction if the group embrace genetic and neuro-explanations, but reject genetic research.

Reductionism

Ortega [18] assesses many of the critiques raised here, and points to the rise of neurological and biologically based explanations for behavior, which replaced the dominant psychoanalytic models of the 1970s. For autism, the "refrigerator mother" theory was interpreted to mean that autism was a reaction to an emotionally deprived upbringing. The shift to a primarily genetic and neurological understandings of autism in the 1980s was ushered in by the pioneering twin studies that provided conclusive evidence of the heritability of autistic traits [22]. The history of autism and how children's mental health and deviance has been variously conceived is covered in many excellent texts, all worth reading [13, 23–25].

There is also bio-medicalization [26]. This process is defined by Conrad as virtually the reverse of neurodiversity: the transformation of everyday human conditions and behaviors into diagnosable, treatable disorders that come to fall under medical jurisdiction [27]. Silverman has cited the increasing diagnosis of autism as an instance of medicalization [13]. Hedgecoe [28] has written about the process of geneticization, through which a condition comes to be understood primarily as genetic. Bumiller has written about this with reference to autism [29].

The rise of neuro-understandings is another example of the way behaviors are now framed as having biological underpinnings. Satel and Lilienfield [30] call this process "Neurocentrism": that is, the tendency to use neurological explanations to explain aspects of a person's behavior, e.g. Shannon Rosa's account of her son: "he was born with his autistic brain" (Chapter 12).

Both neurocentrism and geneticization have been critiqued as forms of biological reductionism, partly because they de-emphasize the complexity through which behaviors are shaped. A reductionist account would see psychological facts as reducible to neurological facts, which is in turn are reducible to biological facts. One consequence of labeling with a diagnostic category like autism is that autistic people's behavior is attributed to brain difference, rather than being under their control [31, 32]. The philosopher of biology Dupré argues against such reductionist accounts of human behavior [33], and opposes the use of causal language.

That genes "cause" autism can be critiqued on many levels. First, behavioral traits which identify autism are exacerbated by the social context and they change over a child's development, so do not imply a fixed state. Some children classified as having autism outgrow their diagnosis, so autism is not necessarily a stable neurological difference throughout the lifecourse [34]. Second, genetic association studies show there is a multi-factorial, complex, genetic predisposition to autism which interacts with epigenetic factors. Third, the environment may alter the development of a person's neurology throughout the lifecourse, whatever their genetics. Fourth, neurology is not fixed but constantly adapted through learning. Dupré argues a better model would be *probabilistic causality*, where nothing "causes" something else but instead increases the chances of it happening.

Although some activists writing in this collection adopted simpler reductionist models, others show their knowledge of complexity: "Genetic research indicates that at least twenty different genes can signal a predisposition to autistic development" (Seidel, Chapter 7); "the genetic factors involved in autism [are] very complex" (Evans, Chapter 9).

Embracing a more nuanced understanding may be a better reflection of reality, but perhaps neglects the impact causal models can have on the real world. For parents of autistic children, for example, the geneticization of autism has meant thousands of mothers escape the guilt and blame that "refrigerator mother" theory engendered. For neurodiversity activists, neuro-models can be a potent instrument to securing accommodations, services, and rights, and gaining political recognition [18].

Medicalization and the Master's House

More widely, medicalization has been defined as the process through which normal behaviors come to fall under medical jurisdiction [27]. In this sense, the neurodiversity movement seeks to de-medicalize autism, because members have argued that autism is a part of normal human variation and should not be considered as a disorder or in medicalized terms. But at the same time, underpinning these arguments are very biologically/neurologically-grounded models of understanding of difference (e.g. "neuro"-differences, "neuro"-diversity). Arguably, these are part of the medicalized framework. The neurodiversity activists therefore co-opt parts of the medical model, whilst espousing broad opposition to the medical model of autism. The whole volume illustrates how a medicalized understanding of autism as a diagnosis of autism spectrum disorder differs from the understanding of autism in the accounts of neurodiversity activists. Contributors describe the medical framework as tending to pathologize people:

> We were so used to being misunderstood, patronized and pathologized (Dekker)
> The dangers of using a selective pathological description (Baggs)
> Undue pathologization of their traits (Seidel)

More general resistance is expressed, e.g. "our community had been segregated by a medical model which insisted on separating us from our natural peers" (Craine, Chapter 19). Such accounts show opposition to medicalization, yet sometimes deploy medicalized rhetoric. In this sense the NDM can be interpreted as using the Master's tools medical narratives] to tear down the Master's house [the medical taxonomic framework], a euphemism used by Lorde [3], a black lesbian feminist.

Lorde's reference to the inability of "the Master's tools to dismantle the Master's house" [3] is a critique of utilizing the rules of those in power. Lorde is arguing that playing their (the Master's) game cannot bring about genuine revolutionary change. ASAN's input into DSM-5, which defines how autism is understood and identified (Chapter 13), was to gain reform not revolution. ASAN chose to engage with the psychiatric system and play the rules—adopting the language of science and using scientific citations to communicate their political points to represent the best interests of the autistic community. Kapp and Ne'eman thus provides an excellent counter to the "Master's tools" argument in Chapter 13, showing how ASAN brought about meaningful change. Their tactic of engagement with the establishment is reminiscent of the AIDS activists' strategy of becoming proficient with and co-opting scientific empiricism to become scientific experts and communicate with the scientific establishment [35]. At the same time, they argue that the DSM should openly acknowledge that diagnosis is partly shaped as a political, as well as a scientific process, a point made by others in the academic literature. Aronowitz, for example, argues that symptoms become "a disease" through social and political processes [36]. Acknowledging and encouraging the socio-political nature of psychiatric taxonomy of DSM underlines that autism is an entity that is *both* constructed *and* is a neurodevelopmental difference.

It can be pathologizing to be given a diagnosis of disorder. At the same time, the medical diagnosis can act as an explanation of the experience of difference, a rallying point for political action, a tool to unlock resources and services, and a first step in moving toward entering a community. Kapp and Ne'eman acknowledge there is a need for a diagnosis but suggest a middle way might be possible: "We believe that identification of autism should transition to a non-pathological system."

I have questions about one's responsibility for behavior and the way diagnosis (or a diagnostic category) excuses deviant (or poor) behavior as a form of sickness [37]. The problem is a "born this way" narrative de-emphasizes personal responsibility, which can be tremendously helpful, but can sometimes be used as an excuse to avoid culpability. If a person was diagnosed with Tourette's, their swearing would be seen as an involuntary aspect of their condition, promoting tolerance and acceptance. On the other hand, what about a psychopathic person who manipulates and exploits others? This behavior also seems to fall under the wider neurodiversity banner, so does the neurodiversity movement require acceptance that the person can't control their behavior, and therefore not culpable or responsible for their behavior? Should neurodivergence be more often considered as a legal mitigating factor? ADHD is strongly associated with criminal behavior, for example [38]. Again, perhaps a more nuanced questioning of repercussions of these concepts may be necessary.

Group Think

A final critique is the accusation that the movement requires conformity. Some complain the movement may engender social conformism through doctrinal thinking that excludes autistic people with diverse viewpoints. Hiari asserts that "The neurodiversity movement epitomizes groupthink" [21], and cites the expulsion of autistics like pro-cure Mitchell and Google engineer Damore (who wrote that male/female disparities can be partly explained by biological difference). I am not sure of the legitimacy of these arguments if Neurodiversity is considered as a political ideology. If it is thought of this way, the neurodiversity movement operates more like a political group, and is entitled to throw out members who express views contrary to the party line. Just because you are female does not make you a feminist.

Conclusion

These are some of the critiques faced by the neurodiversity movement. Whilst the movement seeks a non-pathologizing form of identity and the autistic activist community and allies have made a unique contribution toward this, this aim may sometimes sit uncomfortably with pragmatic forms that their activism takes.

References

1. Haraway, D. (1988). Situated knowledges: The science question in feminism and the privilege of partial perspective. *Feminist Studies, 14*(3), 575–599.
2. Hedges, C. (2018, February 5). The bankruptcy of the American Left. *truthdig*. Retrieved from https://www.truthdig.com.

3. Lorde, A. (1983). The master's tools will never dismantle the master's house. In C. Moraga & G. Anzaldúa (Eds.), *This bridge called my back: Writings by radical women of colour* (pp. 94–101). New York: Kitchen Table Press.
4. Runswick-Cole, K. (2014). 'Us' and 'them': The limits and possibilities of a 'politics of neurodiversity' in neoliberal times. *Disability & Society, 29*(7), 1117–1129.
5. Bailey, A., & Parr, J. (2003). Implications of the broader phenotype for concepts of autism. In G. R. Bock & J. A. Goode (Eds.), *Autism: Neural basis and treatment possibilities* (Vol. 251 of Novartis Foundation Symposia). Chichester, UK: Wiley.
6. Couteur, A., Bailey, A., Goode, S., Pickles, A., Robertson, S., Gottesman, I., et al. (1996). A broader phenotype of autism: The clinical spectrum in twins. *Journal of Child Psychology and Psychiatry, 37*(7), 785–801.
7. Posserud, M.-B., Lundervold, A. J., & Gillberg, C. (2006). Autistic features in a total population of 7–9-year-old children assessed by the ASSQ (Autism Spectrum Screening Questionnaire). *Journal of Child Psychology and Psychiatry, 47*(2), 167–175.
8. Steer, C. D., Golding, J., & Bolton, P. F. (2010). Traits contributing to the autistic spectrum. *PLoS ONE, 5*(9), e12633.
9. Larsson, H., Anckarsater, H., Råstam, M., Chang, Z., & Lichtenstein, P. (2012). Childhood attention-deficit hyperactivity disorder as an extreme of a continuous trait: A quantitative genetic study of 8,500 twin pairs. *Journal of Child Psychology and Psychiatry, 53*(1), 73–80.
10. Maranon, A. (2018, July 31). *[Answer to question:] Why don't neurotypical people understand that many people on the Autism Spectrum (like me) don't know how they feel unless we see an obvious emotional reaction (such as crying)?* Retrieved from https://www.quora.com/Why-dont-neurotypical-people-understand-that-many-people-on-the-Autism-Spectrum-like-me-dont-know-how-they-feel-unless-we-see-an-obvious-emotional-reaction-such-as-crying.
11. Griffin, E. M. (2009). *A first look at communication theory* (7th ed.). New York, NY: McGraw-Hill.
12. Armstrong, T. (2015). The myth of the normal brain: Embracing neurodiversity. *Journal of Ethics, 17*(4), 348–352.
13. Silverman, C. (2011). *Understanding autism: Parents, doctors, and the history of a disorder*. Princeton, NJ: Princeton University Press.
14. Horton, H. (2018, May 22). Labour suspends male activist who stood as women's officer 'because he identifies as a woman on Wednesdays'. *The Telegraph*. Retrieved from https://www.telegraph.co.uk.

15. Mitchell, J. (2019, January 19). The dangers of 'neurodiversity': Why do people want to stop a cure for autism being found? *The Spectator.* Retrieved from https://www.spectator.co.uk.
16. Russell, G., Starr, S., Elphick, C., Rodogno, R., & Singh, I. (2018). Selective patient and public involvement: The promise and perils of pharmaceutical intervention for autism. *Health Expectations, 21*(2), 466–473.
17. Kansen, P. S. (2016, May 25). What the neurodiversity movement gets wrong about autism. *Pacific Standard.* Retrieved February 11, 2019, from https://psmag.com.
18. Ortega, F. (2009). The cerebral subject and the challenge of neurodiversity. *BioSocieties, 4*(04), 425–445.
19. Casanova, M. (2015, January 5). *The neurodiversity movement: Lack of trust* (Web log post). Retrieved February 11, 2019, from https://corticalchauvinism.com/2015/01/05/the-neurodiversity-movement-lack-of-trust/.
20. Jaarsma, P., & Welin, S. (2012). Autism as a natural human variation: Reflections on the claims of the neurodiversity movement. *Health Care Analysis, 20*(1), 20–30.
21. Hiari, T. (2018, April 8). Neurodiversity is dead. Now what? *Mad in America.* Retrieved February 11, 2019, from https://www.madinamerica.com.
22. Freitag, C. M. (2007). The genetics of autistic disorders and its clinical relevance: A review of the literature. *Molecular Psychiatry, 12*(1), 2.
23. Evans, B. (2017). *The metamorphosis of autism: A history of child development in Britain.* Manchester, UK: Manchester University Press.
24. Nadesan, M. (2005). *Constructing autism: Unravelling the 'truth' and understanding the social.* London and New York: Routledge.
25. Silberman, S. (2015). *Neurotribes: The legacy of autism and the future of neurodiversity.* New York, NY: Penguin.
26. Clarke, A. E., Shim, J. K., Mamo, L., Fosket, J. R., & Fishman, J. R. (2003). Biomedicalization: Technoscientific transformations of health, illness, and U.S. biomedicine. *American Sociological Review, 68*(2), 161–194.
27. Conrad, P. (2008). *The medicalization of society: On the transformation of human conditions into treatable disorders.* Baltimore, MD: John Hopkins University Press.
28. Hedgecoe, A. (2001). Schizophrenia and the narrative of enlightened geneticization. *Social Studies of Science, 31*(6), 875–911.
29. Bumiller, K. (2009). The geneticization of autism: From new reproductive technologies to the conception of genetic normalcy. *Signs: Journal of Women in Culture and Society, 34*(4), 875–899.

30. Satel, S., & Lilienfeld, S. O. (2013). *Brainwashed: The seductive appeal of mindless neuroscience.* New York: Basic Civitas Books.
31. Farrugia, D. (2009). Exploring stigma: Medical knowledge and the stigmatisation of parents of children diagnosed with autism spectrum disorder. *Sociology of Health & Illness, 31*(7), 1011–1027.
32. Singh, I. (2013). Brain talk: Power and negotiation in children's discourse about self, brain and behaviour. *Sociology of Health & Illness, 35*(6), 813–827.
33. Dupré, J. (1998). Against reductionist explanations of human behaviour: John Dupré. *Aristotelian Society Supplementary Volume, 72*(1), 153–172.
34. Fein, D., Barton, M., Eigsti, I. M., Kelley, E., Naigles, L., Schultz, R. T., et al. (2013). Optimal outcome in individuals with a history of autism. *Journal of Child Psychology and Psychiatry, 54*(2), 195–205.
35. Epstein, S. (1995). The construction of lay expertise: AIDS activism and the forging of credibility in the reform of clinical trials. *Science Technology Human Values, 20*(4), 408–437.
36. Aronowitz, R. A. (2001). When do symptoms become a disease? *Annals of Internal Medicine, 134*(9 Pt 2), 803–808.
37. Conrad, P., & Schneider, J. W. (1992). *Deviance & medicalization: From badness to sickness.* Philadelphia, PA: Temple University Press.
38. Fletcher, J., & Wolfe, B. (2008). Child mental health and human capital accumulation: The case of ADHD revisited. *Journal of Health Economics, 27*(3), 794–800.

Open Access This chapter is licensed under the terms of the Creative Commons Attribution 4.0 International License (http://creativecommons.org/licenses/by/4.0/), which permits use, sharing, adaptation, distribution and reproduction in any medium or format, as long as you give appropriate credit to the original author(s) and the source, provide a link to the Creative Commons license and indicate if changes were made.

The images or other third party material in this chapter are included in the chapter's Creative Commons license, unless indicated otherwise in a credit line to the material. If material is not included in the chapter's Creative Commons license and your intended use is not permitted by statutory regulation or exceeds the permitted use, you will need to obtain permission directly from the copyright holder.

22

Conclusion

Steven K. Kapp

In less than 30 years of organized activity, the autism rights branch of the neurodiversity movement has progressed from the fringe to the edge of the establishment. As it has matured from a mainly socio-cultural scope to an active part of a cross-disability rights coalition, the neurodiversity movement has shifted increasing focus toward not only what it opposes, but also what it supports. Increasing engagement on practical issues from the balance between safety and autonomy to reproductive and parenting rights have made the boundaries of activists' positions clearer and offered practical support in areas, such as through toolkits and multiple book presses owned by autistic and other neurodivergent people. Autistic people (and our organizations) have become increasingly included and recognized

S. K. Kapp (✉)
College of Social Sciences and International Studies, University of Exeter, Exeter, UK
e-mail: steven.kapp@port.ac.uk

Department of Psychology, University of Portsmouth, Portsmouth, UK in autism advocacy, for example we have been consulted (alongside parent-led organizations) on matters from Hillary Clinton's 2016 presidential campaign [1] to an autistic *Sesame Street* character [2], producing results that pleased the autism community generally.

This concluding chapter will summarize the stories of events told by leaders of the autistic community and neurodiversity movement in this book and provide context to their significance in the broader political and social world. It will also address some of the critiques of the movement and suggest avenues where activists can make further progress.

The Story so Far

The neurodiversity movement has had overwhelmingly positive influence on clinical and scientific directions in the autism field. Interventions and supports have increasingly adopted approaches and goals more aligned with the neurodiversity framework (e.g. building from strengths and interests to develop useful skills rather than normalization; den Houting [3]). A growing number of leading scientists, beginning mainly in the U.K., have become openly interested in the movement and do work with relatively high compatibility with the neurodiversity perspective [4]. Research (e.g. in the U.S.) has moved toward the neurodiversity movement's research priorities, with a much higher proportion of funding awarded for studies on services, adults, and underserved populations [5]. Similarly, studies have increasingly recognized autism's complexity to the point that recognition that traits which may be advantageous have become part of the state of the science, and autistic adults have demonstrated the most expertise in autism according to the latest scientific understanding [6]. Many years after autistic individuals like Temple Grandin (followed by the organized autistic community) influenced the autism field to incorporate atypical reactions to sensory input in the diagnostic criteria, participatory research partnerships with autistic people have finally become a popular trend (especially in the anglophone world; Nicolaidis et al. [7]; Silberman [8]).

Autistic community and neurodiversity movement leaders, especially the contributors to this book, have driven this shift toward inclusion of autistic people and their goals. Autistic voices have raised the profile of harms against autistic and neurodivergent people and a sense of autistic identity, such as Sinclair's essay "Don't Mourn for Us" (see Pripas-Kapit, Chapter 2) and websites or webpages autistics.org (Tisoncik, Chapter 5), Getting the Truth Out (Baggs, Chapter 6), and "The Autistic Genocide Clock" (Evans, Chapter 9). As an autistic-led scholarly journal within critical autism studies, *Autonomy* has helped to preserve key autistic writings like "Don't Mourn for Us" and demonstrate the expertise of autistic people in shaping academic and lay ideas about autism (Arnold, Chapter 15). Autistic-led organizations that meet in cyberspace such as InLv (Dekker, Chapter 3) and physical space such as Autscape (Bucker, Chapter 8) have provided acceptance for fellow autistic people, further building autistic community. Active efforts to include autistic people who share other marginalized identities, such as who have an oppressed gender (daVanport, Chapter 11) or race (Giwa Onaiwu, Chapter 18), have helped advocates represent and strengthen autistic community and activism. Allies such as Seidel of neurodiversity.com (Chapter 7) and the non-autistic editors working alongside autistic editors of The Thinking Person's Guide to Autism (Greenburg and Des Roches Rosa, Chapter 12) have helped the neurodiversity movement gain the credibility, channels, and power to spread the pro-science, pro-autism acceptance agenda to non-autistic relatives, professionals, and researchers. Campaigns led by autistic and other disabled people against medically and legally sanctioned abuses such as chemical restraint through overmedication (Murray, Chapter 4) and institutionalized electric shock therapy (Neumeier and Brown, Chapter 14) have raised awareness of these practices and gathered momentum against them. Meanwhile, organizations and individuals have incorporated the neurodiversity framework into their everyday work outside of formal activism. These include the AASPIRE community-based participatory research project that has attracted federal funding and international acclaim as a model for including lay and scientific autistic people alike in every phase of academic studies (Raymaker, Chapter 10), and Eric Garcia's journalism that positively and accurately publicizes autism and disability in news and analysis (Chapter 17).

Now the neurodiversity movement has arguably arrived at the threshold of the autism establishment. Autistic activists advised the revision of their own diagnosis in the DSM-5, achieving significant success (although short of our goals), but this happened on an ad hoc basis rather than as part of full and systemic inclusion (Kapp and Ne'eman, Chapter 13). The creation of Neurodivergent Labour in connection with the U.K.'s Labour Party (Craine, Chapter 19) and the National Autistic Taskforce as an outgrowth of the National Autism Project (Murray, Chapter 20) have shown an investment in the neurodiversity movement for research, public policy, and practice in the U.K. for autistic people and beyond, although due to their recentness a fuller assessment of their impact awaits. As the movement has become more mainstream activists have tended to maintain their principles, and abandon counterproductive attempts to moderate activists who antagonize autistic people, as Robison did when he resigned from advising Autism Speaks (Robison, Chapter 16). This action propelled that exceptionally powerful organization to begin to make reforms [9], which suggests the movement may become a more dominant force in autism advocacy while staying true to itself.

Inclusion of Autistic People with Higher Support Needs

While the neurodiversity movement has become more representative of autistic people's developmental and cultural diversity through autistic members and both autistic and non-autistic parents or relatives, the most persistent critiques about it tend to claim that it only serves the needs of autistic people with low support needs—sometimes inaccurately and offensively called "high-functioning". As unfortunately this book's design did not enable autistic people lacking verbal fluency to contribute (although it does have a non-speaking contributor, Baggs), I will attempt to address these concerns in this section.

While the autism rights movement has welcomed autistic people regardless of support needs—as well as non-autistic relatives, support people, or friends as allies—from the beginning (Pripas-Kapit, Chapter 2), the fluidity and complexity of autistic people's support needs make classifying them by functioning levels or labels inaccurate. Speech, language, and communication all differ from one another, and verbal tests tend to underestimate the cognitive abilities of autistic people with little expressive language [10–14]. Reasonable accommodations such as allowing extra time [15] and visual supports [16, 17] support many such autistics to reveal their verbal comprehension and cognitive capabilities. Not only do many such individuals perform as "untestable" on standard IQ tests, but they tend to poorly relate to functioning in autistic people generally [18]. Furthermore, the autism field has failed to identify valid subtypes within the autism spectrum, or any consensus on how to measure autism severity or support needs. This contributed to the decision of the DSM-5 workgroup—influenced by the Autistic Self Advocacy Network to oppose the imposition of a severity scale and frame it as about "support needs" to protect access to services (Kapp and Ne'eman, Chapter 13). Many autistics perform well *because* of the social contexts and supports, and struggle when their enabling environments and services disappear (e.g. after leaving high school; Kapp [19]).

These difficulties with conceptualizing and measuring autistic people's developmental diversity include that autistic people typically have uneven skills (American Psychiatric Association [20]; Kapp and Ne'eman, Chapter 13), and large disparities in our cognitive profiles [21]. The same autistic individuals' behavior [22] and perception [23] has demonstrated exceptional variability to the same task or stimuli over time. Even so-called talents or gifts (where present) vary in their presentation as strengths or weaknesses [24], depending on factors such as the social context [25]. Autistic-typical strengths such as pattern recognition tend to exist across the spectrum, including in minimally verbal children classified as "untestable" [10].

Autistic people also tend to gain skills across our lifespans (APA 2013), and the same activists parents might claim as unlike their child may have presented more severely as children. For example, Sinclair, the main "father" of the neurodiversity movement through their work with Autism Network International (see Chapter 2), noted of ANI co-founders "we had all fit descriptions of 'low functioning' autistic people when we were younger" [26]. All had speech delays as children, such as the onset of semi-reliable independent speech at age 12 for Sinclair, yet their access to speech and functioning continued to vary in daily life as adults [27]. Many non-speaking autistic people participate in the neurodiversity movement today, including serving on the board of organizations like the Autistic Self Advocacy Network and Autism National Committee, authoring papers in a special issue on the movement in a major disability studies journal, writing blogs, co-editing or contributing to books, and (co-)directing or participating in documentaries. Many members have significant impairments and support needs, such as long-time activist Cal Montgomery, one of the most widely cited autistic leaders in this book (by Laura Tisoncik [Chapter 5], Mel Baggs [Chapter 6], and Shain Neumeier and Lydia X. Z. Brown [Chapter 14]), who said, "I am incontinent and cannot live alone, cannot bathe myself, etc." (personal communication, February 22, 2019).

From the movement's beginnings, non-autistic family members have advocated alongside autistic activists to fight for the rights of their relatives, many of whom would struggle to engage as social activists even as everyone self-advocates and communicates through behaviour or other forms of communication [28]. For example, disability rights activist Diana Pastora Carson does not identify as neurodivergent but managed her family's successful fight to remove her significantly impaired autistic brother Joaquin (Carson) from an institution, led by Joaquin's expressed desire to leave that she explained. After a life of enforced behavioral compliance training, chemical and physical restraints, and banishment to an institution, Joaquin settled in his own house of his choosing in the community with the full-time support of publicly funded alternating staff, where he has the support of and friendships with neighbors; contributes to events; and walks, runs, and bikes in the peaceful surrounding countryside. He enjoys work, frequent visits from his close relatives, expanded access to communication (through words, typing, and a board), and attending university courses [29].

Nevertheless, movement *activists* arguably do often have skills more developed in key areas than most people in a population, including in communication. The neurodiversity movement is no different in that regard, and all parts of the autism community leave room for improvement in making their organizations, work, and activities more accessible and inclusive. While autistic-led events such as Autreat in the US (Pripas-Kapit, Chapter 2) and Autscape in the UK (Buckle, Chapter 8) have developed tools, activities, and schedules in chosen venues to try to reconcile competing access needs, they have encountered limits to accommodating everyone. While the movement may underrepresent the most profoundly impaired autistic people with the highest support needs, the broader autism community and autism field share that challenge. Autistic people with ID and high "severity" have become increasingly excluded from autism research [30, 31] .

Serving the Interests of Autistic People with Higher Support Needs

The neurodiversity movement seeks to safeguard and fight for the provision of "the full set of human rights" to all neurodivergent people [32], in all major life domains such as "accessing communication, education, employment, competent medical care, the right to make our own decisions and live on our own terms, friendship, romantic relationships and sexuality, freedom from abuse, or the basic premise of our lives being acceptable" [33]. These rights apply equally to people with higher support needs, and those less empowered suffer the most vulnerability to violation of their rights. For example, autistic people with less recognized or reliable communication may risk greater abuse and neglect from so-called caregivers, with less access to legal recourses. While the right of people with disabilities to exist may strike some as uncontroversial, I have observed even disabled activists express understanding of and advocacy for parents who murder their disabled child, with no one but me speaking to the plight of the child, while serving in the capacity of a public U.S.-based body intended to protect the human rights of people with disabilities. Disabled people would not be tortured with electric shock "treatment" in the Judge Rotenberg Center (Neumeier and Brown, Chapter 14) were it not for protection from parents who influence politicians. These sorts of actions have generally not received the same level of priority from parent-led advocacy organizations, if not condoned by them. When people working for those organizations commit abuse, the workers often experience a metaphorical slap on the wrist rather than prosecution, such as recently a relatively modest fine against a "care" home run by the UK's largest autism organization [34]. As autistic activist Crow [35] argued, "we should not have our humanity be devalued on whether or not we are nonverbal or need 24/7 care for rest [sic] of our lives" (pp. 5–6).

Within the autism community autistic adults and the neurodiversity movement place the highest priority on systemic, practical needs such as services, a more immediate and arguably effective focus than the biological and causation studies that consumes most autism research funding [36]. The neurodiversity movement emphasizes increasing literacy and access to reliable communication, as declared by Ne'eman [37], who implemented this priority while serving on the U.S.'s steering body for funding autism research through successfully pushing for research on augmentative and alternative communication. This is echoed in the U.K. where the members of the National Autistic Taskforce exercise their verbal and other privileges to prioritize the needs of more disabled autistic people. A foundational focus on policy for the most politically mobilized autistic people, like the broader disability rights movement, has helped preserve access to hard-won rights and services while extending others. In the U.S., examples include protecting healthcare and community living through developmental disability services, while raising the minimum wage for workers with disabilities (previously paid as little as pennies on the hour) to parity with the raise for other federally contracted employees [33]. A stronger collaboration in the autism community on supporting the right (and funding) for autistic people to live in their own place with the support needed would enable families to not need to serve as primary caregivers for autistic adults, or at least shifting greater priority to family services could provide relief for familial caregivers.

Parental acceptance of their child's autism helps the parents' understanding of their child, well-being, and the parent–child relationship. This parental acceptance of autism does not relate to the "severity" of the child's externally measured "symptoms" (e.g. by trained observers based on coding of elicited behaviors and semi-structured interactions), but only relates to (fewer or less pronounced traits) according to parental self-report [19]. These findings suggest the attitudes against the neurodiversity movement for autistic people with higher support needs only hold for *subjective* perceptions of autistic people's differences and impairments.

Similarly, external and self-acceptance of autism helps support autistic adults' well-being [38], as it likely does for younger autistic people. While the autism research field lacks analysis of autistic people's views on the core claims of the neurodiversity movement according to their supposed level of support needs, autistic people's functioning has a complex relationship with their core traits and abilities, and one should not make the dangerous assumption that more impaired individuals would more likely oppose the neurodiversity framework. While autistic people with higher support needs undoubtedly face greater risks of denial of basic rights such as autonomy and inclusion, research indicates that autistic people with subtler manifestations of autism and higher cognitive abilities experience more peer bullying, distress, internalized ableism, and exclusion from services [19].

Furthermore, while a study reporting that the social factors related to discrimination and stigma accounted for 72% of the distress experienced by autistic adults had a highly verbal sample [39], this may apply across the autism spectrum. For example, statistical studies have failed to explain self-injury, with little to no relationship to IQ and even anxiety [40]; review and research by Dempsey et al. [41]. Yet this may result from the studies' reliance on parent report; only autistic people have direct access to our emotions, and reporting on autistic children's anxiety has fared better by self-report than parent report [42]. Aggressive behaviors (including against the self) may stem largely from failure of the social environment to meet autistic people's needs, as autistic neurodiversity activist Ballou [43] argues.

Final Thoughts

This book has attempted to document the actions of leading autistic activists in the neurodiversity movement, covering the history at a time when it has undergone different waves in its development, yet not too late to attract most leaders from the countries where it has become most established. It has also sought to explain the concepts of *neurodiversity* and the beliefs and work of the neurodiversity *movement*, engaging with critiques at a time when misunderstandings linger. "Neurodiversity-lite" has seeped into autism culture (adopting some of the rhetoric of the movement but not truly implementing the principles: Neumeier [44]), perhaps mainly due to ignorance but also co-option of the movement's strengthening force (e.g. changing an organization's name but not its practices—"On Autism Orgs" [45]). The movement has made great progress and has begun to enter politics, yet unless the movement further coalesces in a broader coalition in more regions of the globe, its impact on combating the growing austerity in a global competitive economy may be limited. Future books and works of scholarship and activism may further shed light on the current status of the movement beyond its origins (beyond autism and well beyond mainly anglophone countries), and deconstruct paths forward for helping neurodivergent people receive the support and respect we need.

References

1. Gearan, A., & Bernstein, L. (2016, January 5). Hillary Clinton outlines autism proposal, calls for nationwide early screening initiative. *The Washington Post*. Retrieved from https://www.washingtonpost.com/news/post-politics/wp/2016/01/05/hillary-clinton-outlines-autism-proposal-calling-for-nationwide-early-screening-initiative/.
2. Willingham, E. (2015, October 22). 'Sesame Street' introduces autistic girl into the neighborhood. *Forbes*. Retrieved from http://www.forbes.com/sites/emilywillingham/2015/10/22/sesame-street-introduces-autistic-girl-into-the-neigborhood/#2715e4857a0b7afe36e4855a.
3. den Houting, J. (2019). Neurodiversity: An insider's perspective. *Autism, 23*(2), 271–273.
4. Pellicano, E., & Stears, M. (2011). Bridging autism, science and society: Moving toward an ethically informed approach to autism research. *Autism Research, 4*(4), 271–282.
5. Pellicano, L., Mandy, W., Bölte, S., Stahmer, A., Lounds Taylor, J., & Mandell, D. S. (2018). A new era for autism research, and for our journal. *Autism, 22*, 82–83.
6. Gillespie-Lynch, K., Kapp, S. K., Brooks, P. J., Pickens, J., & Schwartzman, B. (2017). Whose expertise is it? Evidence for autistic adults as critical autism experts. *Frontiers in Psychology, 8*, 438.
7. Nicolaidis, C., Raymaker, D., Kapp, S. K., Baggs, A., Ashkenazy, E., McDonald, K., et al. (in press). Practice-based guidelines for the inclusion of autistic

adults in research as co-researchers and study participants. *Autism.* Advance online publication. https://doi.org/10.1177/1362361319830523.
8. Silberman, S. (2017). Beyond "Deficit-Based" thinking in autism research. Comment on "Implications of the idea of neurodiversity for understanding the origins of developmental disorders" by Nobuo Masataka. *Physics of Life Reviews, 20,* 119–121.
9. Rosenblatt, A. (2018). Autism, advocacy organizations, and past injustice. *Disability Studies Quarterly, 38*(4). Retrieved from http://dsq-sds.org/article/view/6222/5137.
10. Courchesne, V., Meilleur, A. A. S., Poulin-Lord, M. P., Dawson, M., & Soulières, I. (2015). Autistic children at risk of being underestimated: School-based pilot study of a strength-informed assessment. *Molecular Autism, 6*(1), 12.
11. Grondhuis, S. N., & Mulick, J. A. (2013). Comparison of the Leiter International Performance Scale-Revised and the Stanford-Binet Intelligence Scales, in children with autism spectrum disorders. *American Journal on Intellectual and Developmental Disabilities, 118*(1), 44–54.
12. Krueger, K. K. (2013). *Minimally verbal school-aged children with autism: Communication, academic engagement and classroom quality* (Doctoral dissertation). Retrieved from the eScholarship UCLA Electronic Theses and Dissertations database (UMI No. 11894).
13. Shah, A., & Holmes, N. (1985). Brief report: The use of the Leiter International Performance Scale with autistic children. *Journal of Autism and Developmental Disorders, 15*(2), 195–203.
14. Tsatsanis, K. D., Dartnall, N., Cicchetti, D., Sparrow, S. S., Klin, A., & Volkmar, F. R. (2003). Concurrent validity and classification accuracy of the Leiter and Leiter-R in low-functioning children with autism. *Journal of Autism and Developmental Disorders, 33*(1), 23–30.
15. McGonigle-Chalmers, M., & McSweeney, M. (2013). The role of timing in testing nonverbal IQ in children with ASD. *Journal of Autism and Developmental Disorders, 43*(1), 80–90.
16. McGonigle-Chalmers, M., Alderson-Day, B., Fleming, J., & Monsen, K. (2013). Profound expressive language impairment in low functioning children with autism: An investigation of syntactic awareness using a computerised learning task. *Journal of Autism and Developmental Disorders, 43*(9), 2062–2081.
17. Mucchetti, C. A. (2013). Adapted shared reading at school for minimally verbal students with autism. *Autism, 17*(3), 358–372.

18. Rudacille, D. (2011, January 6). *IQ scores not a good measure of function in autism*. Simons Foundation Autism Research Initiative. Retrieved from http://sfari.org/news-and-opinion/.
19. Kapp, S. K. (2018). Social support, well-being, and quality of life among individuals on the autism spectrum. *Pediatrics, 141*(Suppl. 4), S362–S368.
20. American Psychiatric Association. (2013). *Diagnostic and Statistical Manual of Mental Disorders (DSM-5®)*. Washington, DC: Author.
21. Jones, C. R., Happé, F., Golden, H., Marsden, A. J., Tregay, J., Simonoff, E., et al. (2009). Reading and arithmetic in adolescents with autism spectrum disorders: Peaks and dips in attainment. *Neuropsychology, 23*(6), 718.
22. Geurts, H. M., Grasman, R. P., Verté, S., Oosterlaan, J., Roeyers, H., van Kammen, S. M., et al. (2008). Intra-individual variability in ADHD, autism spectrum disorders and Tourette's syndrome. *Neuropsychologia, 46*(13), 3030–3041.
23. Haigh, S. M. (2018). Variable sensory perception in autism. *European Journal of Neuroscience, 47*(6), 602–609.
24. Happé, F. (2018). Why are savant skills and special talents associated with autism? *World Psychiatry, 17*(3), 280–281.
25. Russell, G., Kapp, S. K., Elliott, D., Elphick, C., Gwernan-Jones, R., & Owens, C. (2019). Mapping the autistic advantage from the experience of adults diagnosed with autism: A qualitative study. *Autism in Adulthood, 1,* 124–133 (Advance online publication).
26. Sinclair, J. (2012). Autism Network International: The development of a community and its culture. In J. Bascom (Ed.), *Loud hands: Autistic people, speaking* (pp. 17–48). Washington, DC: The Autistic Press. Retrieved from http://www.autreat.com/History_of_ANI.html.
27. Silberman, S. (2015). *NeuroTribes: The legacy of autism and the future of neurodiversity*. New York: Avery.
28. Baggs, A. M. (2012). The meaning of self advocacy. In J. Bascom (Ed.), *Loud hands: Autistic people, speaking* (pp. 315–319). Washington, DC: The Autistic Press.
29. Carson, D. (2017). Walking with Joaquin. *TEDxTalks* (Video file). Retrieved from https://www.youtube.com/watch?v=ruXB3lbiD3U.
30. Russell, G., Mandy, W., Elliott, D., White, R., Pittwood, T., & Ford, T. (2019). Selection bias on intellectual ability in autism research: A cross-sectional review and meta-analysis. *Molecular Autism, 10*(1), 9.
31. Stedman, A., Taylor, B., Erard, M., Peura, C., & Siegel, M. (2018). Are children severely affected by autism spectrum disorder underrepresented in

treatment studies? An analysis of the literature. *Journal of Autism and Developmental Disorders* (Advance online publication). https://doi.org/10.1007/s10803%96018%2D3844%2Dy.

32. Ryskamp, D. A. (2016). *I'm a pro-neurodiversity advocate. Here's what our critics never get right but don't seem to care about, either.* Retrieved from https://autisticacademic.com/2016/05/26/im-a-pro-neurodiversity-advocate-heres-what-our-critics-never-get-right-but-dont-bother-to-correct-either/.

33. Ballou, E. P. (chavisory). (2018, February 6). *What the neurodiversity movement does—And doesn't—Offer* (Blog post). Retrieved from http://www.thinkingautismguide.com/2018/02/what-neurodiversity-movement-doesand.html.

34. Morris, S., & Brindle, D. (2019, March 10). CQC rebuked for failure to prosecute charity over care home abuse. *The Guardian.* Retrieved from https://www.theguardian.com/society/2019/mar/10/cqc-rebuked-for-failure-to-prosecute-national-autistic-society-over-mendip-house-care-home-abuse.

35. Crow, M. (2017). *Anarchism: In the conversations of neurodiversity.* The Anarchist Library.

36. Pellicano, E., Dinsmore, A., & Charman, T. (2014). What should autism research focus upon? Community views and priorities from the United Kingdom. *Autism, 18*(7), 756–770.

37. Ne'eman, A. (2010). The future (and the past) of autism advocacy, or why the ASA's magazine, *The Advocate*, wouldn't publish this piece. *Disability Studies Quarterly, 30*(1). Retrieved from http://dsq-sds.org/article/view/1059/1244.

38. Cage, E., Di Monaco, J., & Newell, V. (2018). Experiences of autism acceptance and mental health in autistic adults. *Journal of Autism and Developmental Disorders, 48*(2), 473–484.

39. Botha, M., & Frost, D. M. (2018). Extending the minority stress model to understand mental health problems experienced by the autistic population. *Society and Mental Health* (Advance online publication). https://doi.org/10.1177/2156869318804297.

40. d'Arc, B. F., Dawson, M., Soulières, I., & Mottron, L. (2012). Self-injury in autism is largely unexplained: Now what? *Journal of Autism and Developmental Disorders, 42*(11), 2513–2514.

41. Dempsey, J., Dempsey, A. G., Guffey, D., Minard, C. G., & Goin-Kochel, R. P. (2016). Brief report: Further examination of self-injurious behaviors in children and adolescents with autism spectrum disorders. *Journal of Autism and Developmental Disorders, 46*(5), 1872–1879.

42. White, S. W., Lerner, M. D., McLeod, B. D., Wood, J. J., Ginsburg, G. S., Kerns, C., et al. (2015). Anxiety in youth with and without autism spectrum disorder: Examination of factorial equivalence. *Behavior Therapy, 46*(1), 40–53.
43. Ballou, E. P. (chavisory). (2014, May 11). *A checklist for identifying sources of aggression* (Blog post). Retrieved from http://wearelikeyourchild.blogspot.com/2014/05/a-checklist-for-identifying-sources-of.html.
44. Neumeier, S. M. (2018). *'To Siri with Love' and the problem with neurodiversity lite*. Available at https://rewire.news/article/2018/02/09/siri-love-problem-neurodiversity-lite/.
45. On Autism Orgs. (2019, January 28). Retrieved from www.thinkingautismguide.com/p/position.html.

Open Access This chapter is licensed under the terms of the Creative Commons Attribution 4.0 International License (http://creativecommons.org/licenses/by/4.0/), which permits use, sharing, adaptation, distribution and reproduction in any medium or format, as long as you give appropriate credit to the original author(s) and the source, provide a link to the Creative Commons license and indicate if changes were made.

The images or other third party material in this chapter are included in the chapter's Creative Commons license, unless indicated otherwise in a credit line to the material. If material is not included in the chapter's Creative Commons license and your intended use is not permitted by statutory regulation or exceeds the permitted use, you will need to obtain permission directly from the copyright holder.

Index

A

AAP (Autistic Advisory Panel), National Autism Project 281, 282
AASPIRE (Academic Autism Spectrum Partnership in Research and Education) 13, 133–142, 307
ABA (applied behavioral analysis) ix, 158, 273, 278
ableism 67, 152, 160, 197, 250, 313
accessibility 114
activism 5, 10, 14, 148, 266, 314
  autistic, 46, 75, 116, 131, 133, 134–141, 307
  and autistics.org, 65, 67, 68
  and disability rights, v
  and electric shock "treatment", 201, 203
  intersectional, 152
  and Judge Rotenberg Center, 204
  neurodiversity, 35, 134–141, 155, 294, 300, 305
  political, 65, 72
  scientific, 138
  social justice, 150
adaptive functioning 7, 185, 187
ADAPT, National 199
ADHD (Attention-Deficit Hyperactivity Disorder) 259, 263, 272, 288, 299
ADH-L (Academic Dick Heads List) 78
advocacy
  autism. *See* autism advocacy
  neurodiversity 155–165
AFF (Aspies for Freedom) 126–127, 128, 131, 201

All the Weight of Our Dreams: On Living Racialized Autism 243, 250
American Psychiatric Association 9. *See also* diagnosis/Diagnostic and Statistical Manual (DSM)
Americans with Disabilities Acts (1990 & 2008) 151, 172–173, 198, 235
Ammerdown Centre 110
Andrews, David 54, 215
ANI (Autism Network International) 2, 23, 42, 109, 136, 216, 309
"An Anthropologist on Mars" 89
"anti-cure" perspective on autism 293
anti-Judge Rotenberg Center activism 204
antisocial behavior 117
APANA (Autistic People Against Neuroleptic Abuse) 12, 51–61, 278
APPGA (All-Party Parliamentary Group on Autism) 57–58, 277
ARGH (Autism Rights Group Highlands) 284
ASAN (Autistic Self Advocacy Network). *See* Autistic Self Advocacy Network (ASAN)
"ASD in DSM-5: What the Research Shows and Recommendations for Change" 181
Ashkenazy, E. 243, 244
Ashton, Karen 59
"Ask a Neurotypical" panel, at Autreat 35
#asperger Internet Relay Chat channel 47
Asperger's Syndrome 46, 124, 169, 214, 216, 233, 294

Aspergia 125–126, 131
"Aspie Supremacism" 171
assault, medical treatment as 59. *See also* electric shock "treatment"
AutAdvo 92, 93
AutCom 97
#autfriends/#autism 70–73
autism
  diagnostic criteria for 167–189, 306
  politics of, 237
  as response to trauma, 188
  severity scale for diagnosis of, 167, 175, 177, 179, 183–185, 183–184, 187, 309
  single unified diagnosis of, 169, 171, 180
  as social communication disability, 249
  and vaccines. *See* vaccines, and autism
Autism Acceptance Day/Autism Acceptance Month 12
Autism Act (2009) (UK) 15
autism advocacy 11, 14, 24, 98, 157, 306, 308
  and Autism Speaks, 221, 228, 230
AutismAndRace.com 250
autism-as-tragedy paradigm 26–27, 28, 30, 33, 34, 91, 229
Autism Awareness Day 239
autism diagnosis 23, 90, 167–189, 263, 296
Autism Dividend, The 282
"autism epidemic" 92, 94, 223, 238
Autism Every Day 95, 130, 148
Autism Hub 97, 126, 129–131
Autism in Adulthood 140

Autism/Neurodiversity Manifesto 3, 15, 255–274
autism policy 221
autism quackery 156
autism research 151, 224, 229, 311, 312
   and dehumanization, 73
   ethical concerns with, 7
   participatory, 13, 133, 139
   problematic, 135
autism rights v, 1, 2, 8, 14, 305, 308
Autism Society of America 12, 83, 95, 180
Autism Speaks 130, 131, 221–231, 278, 279
autism spectrum 9, 169, 170, 180, 185, 248, 298
autism therapy ix, 6, 60
autism-vaccine disinformation 158
autism-vaccine litigation 96
Autistica 277–281
autistic activism 46, 75, 116, 131, 134–141, 307
autistic culture 1, 35–37
"Autistic Distinction, The" 92–93
Autistic Genocide Clock v, 13, 123–131, 307
autistic identity 134, 158, 201, 306
Autistic Passing Project 174
Autistic Pride Day 12
Autistic Self Advocacy Network (ASAN) 9, 130–131, 136, 148–150, 158, 231
   DSM-5 lobbying/recommendations by, 167–189, 309
autistic space 4, 13, 35, 111–114, 117, 118, 120, 121

autistic spectrum (InLv), independent living on the 41–48, 109, 110
Autistics Speaking Day 12
"autistic superiority" 73, 289
autistics.org 12, 65–75, 134, 295, 307
autistic traits 26, 70, 91, 120, 126, 288, 296
   and DSM-5, 172, 176, 185, 187, 189
autoethnography 216
autonomy 2, 8, 58, 198, 206, 246, 277, 305, 313
Autonomy (journal) 14, 211–219, 307
Autreat 12, 13, 24, 30, 35, 36, 48, 97, 216, 310
   and Autscape, 109, 110, 112–114
Autscape 13, 48, 109–121, 282, 292, 295, 307, 310
aversives 200, 203, 205, 206
"awareness" 83, 198, 238
AWN (Autistic Women and Non-Binary Network) 13, 147–154, 157

B

Baggs, Mel 57, 98, 158, 198, 295, 298
Bartak, Lawrence 27
BBS (bulletin board systems) 41, 43, 44
behaviorism 58, 175, 195–196, 198, 279
Berkowitz, Lori 151, 246
Bettelheim, Bruno 68, 68n2
biology 3, 5, 46, 225, 228, 288, 297
bio-medicalization 296

biomedical model, Western 8
Birmingham, University of 57, 60, 215, 218, 280
blogging 129, 134, 161, 201, 258–260
Blume, Harvey 45, 47, 90
Booth, Janine 268–270, 274
Bovell, Virginia 58, 277
Bowen, Paul 59, 60
Breakstone, Savannah Logsdon 150, 151
Bridging the Gaps 31, 32
broad autism phenotype/broader autistic phenotype 91, 288
broader autism community vii, 15, 311
Brook, Kabie 283, 284
Brown, Lydia X.Z. 243, 310
bullying 67, 69, 70, 119, 313
Burns, Charles 110
"butterfly effect" 268–269

C

CAFETY (Community Alliance for the Ethical Treatment of Youth) 202
Cal (Cal Montgomery) 65, 73, 78, 83, 198, 199, 295, 310
CARD (Center for Autism and Related Disorders) ix
caregivers/caregiving 6, 311, 312
Carley, Michael John 170, 181
Carlock, William 26
CAS (Critical Autism Studies) 9, 211, 219, 307
catatonia 42, 47, 52, 55
causation, of autism 25, 26, 95, 149, 155, 238, 312

environmental, 3, 6, 8, 96
CBPR (Community Based Participatory Research) 134–137, 140
"challenging behaviors" 52
chemical restraints 237, 307
chronic illnesses 3
civil rights 41, 126, 127, 129, 134, 136, 234
civil rights movements 10, 66, 72, 147, 152
class 3, 5, 67, 167, 171, 187, 197
clinical practice 183, 189
Clinton, Hillary 237–239, 306
cognitive abilities/cognitive capabilities 185, 309, 313
cognitive demands 172, 178
cognitive differences 29, 92, 93, 99, 309
cognitive impacts 172, 178
cognitive profiles 309
cognitive variety 29, 92, 93, 99, 309
Colley, Mary 214
communication 3, 11, 26, 61, 98. *See also* Social Communication Disorder (SCD)
 assistive, 97
 nonverbal, 91
 social, 183, 185, 186
Communication Shutdown 12
communication support 280, 282, 312
compensatory model 10
compliance training 198, 310
conformism, social and political 47, 300
context-dependence, of autistic functioning 184
co-occurring conditions 9, 259, 261

Cook, Ian 203–204
coping mechanisms/coping skills/coping strategies 6, 8, 172, 177, 178
Corbyn, Jeremy (MP) 266–267, 274
cross-disability matters 1, 2, 4, 13, 131, 200, 305
Cure Autism Now ix
cure movement, of autism ix–x, 8. *See also* "anti-cure" perspective on autism; "pro-cure" perspective on autism

D

Dalmayne, Emma 265
DANDA (Developmental Adult Neuro-Diversity Association) 214
Dawson, Geraldine 224–226, 228
Dawson, Michelle 83, 90, 95, 98, 295
DCD (Developmental Coordination Disorder) 258, 259, 262–264
deficit model, of disability. *See* medical model, of disability
deinstitutionalization 195
Dekker, Martijn 2, 12, 109, 298, 307
dementia 292
Derber, Charles 287
Des Roches Rosa, Shannon 155, 162, 294
desegregation 66
developmental disability (DD) 11, 65, 83, 85, 180, 312
developmental diversity 309
diagnosis/Diagnostic and Statistical Manual (DSM) 9, 23, 42, 90, 167–189, 263, 308

and critiques of the neurodiversity movement, 295, 298, 299
disability
developmental 11, 65, 82, 83, 85, 180, 312
medical model of. *See* medical model, of disability
social model of, 7, 11, 32, 125, 212, 217, 270, 272
Disability Community Day of Mourning 12
disability rights 3, 13, 127, 130, 196, 199, 203, 212
and Autistic Women and Non-Binary Network, 150–152
disability rights activism v
disability rights motto 2, 187, 214, 217
disability rights movements ix, 3, 7, 11, 68, 85, 141, 234, 312
and Autistic Women and Non-Binary Network, 152
disability rights paradigm 136
disability rights principle 2, 187, 214, 217
disability studies 3, 11, 46, 211
discrimination 8, 15, 95, 153, 180, 200, 313
and Autism/Neurodiversity Manifesto, 268, 272
and critiques of the neurodiversity movement, 290, 291
dispersed venues 114
Divergent: when disability and feminism collide 152
"Don't Mourn for Us" 12, 23–37, 219, 307
"Don't Speak for Us" 130

DSM (Diagnostic and Statistical Manual) 185–189, 216
Du Bois, Terri 204
Dunn, Dr. Yo 119, 281, 282–283
Durbin-Westby, Paula 174
dyslexia 257–258, 259, 261, 263, 265
dyspraxia 258, 259, 263, 293
Dyspraxia Foundation 214

E

electric shock "treatment" 196–197, 205, 307, 311
emancipatory research 135
Emergence: Labeled Autistic 24–25, 27–30
empathy 43, 45, 148, 174, 294
empowerment 11, 97, 130, 148, 261, 311
  and AASPIRE, 134, 136, 137, 140, 141
"entrapment" paradigm 28, 35
environmental causation, of autism 3, 6, 8, 96
environmental toxins, as risk factors for autism 6
"epidemic, autism" 92, 94, 223, 238
epigenetics 297
ethics 97, 217
eugenics 7, 13, 125, 128, 131, 278
"Everybody Gets Paid" principle" 247–248
exclusion/exclusivity 12, 41–48, 93, 117, 125, 216, 293, 313
Expanding the Promise for Individuals with Autism Act 237

Exploring Diagnosis: Autism and the Neurodiversity Movement vii–ix

F

face blindness 45
Facebook 258–259, 271
Fair Housing Campaign 65–67
FDA (Food and Drug administration) 73, 203, 205
female genital mutilation (FGM) 290
Fidonet 41, 44
"For The Many Not The Few" 272
"foundation documents" 217
Frith, Uta 215
Fund for Community Reparations for Autistic People of Color's Interdependence, Survival, and Empowerment 251
fundraising 5, 115, 224, 229, 243–244

G

Geier, Mark and David 95, 97
gender/gender identity 5, 112, 134, 151, 152, 243, 292, 307
  and DSM-5, 168, 171, 174, 187
Generation Rescue ix
genetics 93, 127, 225, 228, 296, 297
Gernsbacher, Morton Ann 93
"Getting the Truth Out" website 12, 83, 85, 307
"Getting the Word Out" 12, 83
Giggleswick School 113
Gilfoy, Hilary 278, 280
Grandin, Temple 23, 24–29, 30, 34, 306
Grant, Kathy Xenia 30, 214

GRASP (Global and Regional Asperger Syndrome Partnership) 136, 170, 181
Greenburg, Carol 157, 159, 289
Gross, Zoe 12, 174
group think 300

H

Haley, Boyd 93, 96
Hatch, Senator Orrin 235
health and safety 55, 61
Hiari 294, 300
"high-functioning" autism 184, 294, 308
Hippocratic Oath 59
human rights 1, 4, 59, 82, 134, 164, 311
    and electric shock "treatment", 202, 203

I

IACC (Interagency Autism Coordinating Committee) 95, 139, 140, 175, 230
ideasthesia 147
identity politics 288–290, 291
I Am Autism 131, 226, 227
impairments 7, 114, 118, 163, 261, 293, 310, 312
    and origins of neurodiversity, 28, 32, 35
inclusion/inclusivity viii, 2, 4, 7, 13, 306, 308–311, 313
    and AASPIRE, 137
    and Autism/Neurodiversity Manifesto, 255
    and Autistic Genocide Clock, 131

Autistic Women and Non-Binary Network (AWN), 150
    and Autscape, 109, 114, 115, 117
    and Independent Living on the Autistic spectrum, 46, 47
    and neurodiversity movement, 292, 293, 295
independent living 85, 224
Independent Living on the Autistic Spectrum (InLv) 41–48, 110
individual model, of disability. *See* medical model, of disability
injustice 8, 12, 69, 71–72, 149, 266
"In My Language" 85, 98, 158
INSAR (International Society for Autism Research) 229
institutionalization 78, 93, 135, 195–207, 307
institutions, mental 77, 78
intellectual disabilities 5, 32, 60, 141, 292, 295
interaction badge system 112
Internet 41–43, 45, 258–260, 266
intersectional activism 152
intersectionality 4, 251
intervention ix, 7, 8, 57, 93, 127, 206, 278, 306
    and DSM-5, 175, 185, 189
IRC (Internet Relay Chat) 47, 69, 70, 109
ISNT (Institute for the Study of the Neurologically Typical) v, 12, 73, 90
isolation 26, 27, 31, 71, 169, 197, 198, 262
Israel, Matthew 196, 202
"It's Not a Term Paper" principle 248–249

J

Jordan, Rita 57
Judge Rotenberg Center (JRC) 3, 72, 195–197, 311
justice, social 4, 10, 35, 134, 137, 141, 150, 212

K

Kapp, Steven 168, 299
King, Jr., Martin Luther 65–67

L

Labour, Neurodivergent 271–274, 308
Labour Party (UK) 15, 116, 255, 266, 270, 271–274, 308
Lawson, Wenn 54, 57, 277
learning disabilities 52, 60
legal protections 3, 168, 169, 171–173, 183, 188, 291
Leitch, Kevin 97, 126
Letter from a Birmingham Jail 72
lived experience, of autism 6, 31, 33, 46, 233, 246, 270
Look Me in the Eye 221, 223
losing, but having important effects 77–85
Louder Than Words 223
Lovaas Institute/Lovaas, Ole Ivar ix, 91, 189
"low-functioning" autism 83, 309
"Lupron protocol"/Lupron therapy 7, 95, 96

M

mad pride movement 4
malpractice 89, 202
marginalization 5, 91, 251
"masking" 6, 162–163, 172, 177, 178, 184, 256
"Master's house"/"Master's tools" 288, 298–299
MCA (Mental Capacity Act (2007)) 60
McCollins, Andre/McCollins, Cheryl 196, 202, 203
McDonnell, John 268–274
medical experimentation 197
medicalization/medication 8, 9, 82, 296, 298–300
 and neuroleptic abuse, 51, 52, 54, 56, 60–61
medical model, of disability 1, 7–10, 188, 211, 278
 and Autism/Neurodiversity Manifesto, 259, 261
 and critiques of the neurodiversity movement, 292, 298
medical treatment 59
mental illness 61, 199, 292
Miller, Gregory 202
Milton, Damian 218, 271, 280, 282, 283
misinformation
 about autism 89, 97, 128, 155, 158
 about disability, 158
 and exclusivity, 117, 293
mission statements 91, 138, 226
mistreatment 2, 7, 80, 81
mobilization, of the autistic/neurodivergent community viii, 10, 295
mother-blame 11
movement disorders 51, 52

MR (mental retardation) 29, 32, 79, 80
Msumba, Jennifer 203, 204
Murray, Dinah 12, 218
Myers, Jennifer Byde 156, 162

N

National Autism Project (UK) 15, 280–283, 308
National Autistic Taskforce (UK) 15, 277, 282–283, 308, 312
National Insurance Disability Scheme (Australia) 185
National Journal 233, 235, 236
Ne'eman, Ari 36, 98, 129, 131, 151, 238, 299, 312
neglect 80, 196, 222, 311
Nelson, Amy and Gareth 12, 126
neurocentrism 296
Neurodevelopmental Disorders Workgroup 168, 170, 182
neurodivergences 3, 8, 10, 15, 159, 292, 299
neurodiversity
  origin of term 47
  philosophy of, 35
  politics of, 233–239
neurodiversity activism 35, 133–141, 155, 294, 300
neurodiversity advocacy 155–165
neurodiversity.com 12, 89–99, 307
neurodiversity paradigm 47, 261
Neurodiversity Weblog 95–99
"neurodiversity-lite" 313
neuroleptic abuse 12, 51–61
neurological disabilities 4, 12, 261
NeuroTribes 156
neurotypical, meaning of 12

neurotypical privilege 3, 290
Nicolaidis, Christina 134
Nobody, Nowhere 25, 27–29
"Nothing About Us Without Us" principle v, 2, 158, 187, 246–247, 269, 292
  and Autistic Self Advocacy Network (ASAN), 130
  and *Autonomy* (journal), 215, 218

O

Oliver, Mike 7
Open Journal Systems platform 217
oppression 4, 70, 135, 137, 200, 251
OSR (industrial chelator) 96, 97n2
"Other Half, The" 282
overmedication 4, 307
"own world" paradigm 27, 29, 34

P

paradigm
  autism-as-tragedy 26–28, 30, 33, 34, 91, 229
  disability rights, 136
  "entrapment", 28, 35
  neurodiversity, 47, 261
  "own world", 27, 29, 34
  pathology, 261
  "trapped", 28, 35
Paradiz, Valerie 230
parental grief/parental mourning 30, 33, 34. *See also* "Don't Mourn for Us"
parenting practices, positive 6

participatory research 137, 141, 306, 307. *See also* CBPR (Community Based Participatory Research)
"passing" 172
"Past, Present, and Future" 83, 85
pathologization 12, 99, 298
pathology paradigm 261
personal experience, of autism 6, 31, 33, 46, 233, 247, 270, 290, 295
personal responsibility 288, 299
person-first language 37, 260, 261
"Petition to Defend the Dignity of Autistic Citizens" 94
physical restraint 4, 310
PoC (person of color) 246–248, 250
political activism 65, 72
politics, of autism and neurodiversity 233–239
polypharmacy 52
"Positive Behaviour Support" 60
prejudice 96, 99, 117, 290, 293
prenatal testing, for autism 127–128, 131
"pressure points" analysis 173
primary sensory disabilities 3
privilege 25, 27, 67, 72, 74, 165, 247, 312
 neurotypical, 3, 290
"pro-acceptance" 5, 12
probabilistic causality 297
"pro-cure" perspective on autism 5, 9, 300
proprioception 162–163
prosopagnosia 45
psychiatric matters 3, 54, 65, 199, 201, 292, 298
psychoanalytic child psychology 11

psychogenic theories 6
psychotropics 8, 52, 54, 59, 60
PTSD (Post-Traumatic Stress Disorder) 149, 158, 204
Public Knowledge Project 217
Pukki, Heta 110, 215

Q

quality of life 7, 55, 95, 172, 185, 294

R

race 4, 135, 152, 167, 171, 174, 307
 and first autism and race anthology, 248, 249, 251
racism 67, 197–199, 250
"Real Transparency" principle 249
reasonable accommodations/reasonable adjustments 8, 178, 180, 309
"recovery" 25, 172, 176, 183. *See also Emergence: Labeled Autistic*
reductionism 288, 296–297
"refrigerator mother" theory 29, 296, 297
"regression", into autism 26
rejection 28, 43, 198
Rimland, Bernard 25, 27, 31
risk aversion, "health and safety" 55
risk factors, for autism 6
Robertson, Scott ix, 136, 174
Roll Call 236, 237
Runswick-Cole, Katherine 288, 290

S

Sacks, Oliver 89

Saint John's Autism Listserv 78
Schwarz, Phil 75, 92, 95
scientific activism 138
sedation 52, 53
self-acceptance 23, 30, 312
self-advocacy 5, 10, 85, 131, 148, 164, 180, 238, 310. *See also* Independent Living on the Autistic Spectrum (InLv)
 and AASPIRE, 138, 140
 and electric shock "treatment", 196, 199
self-definition 292
self-determination, right to 4, 7–8, 142, 207
self-diagnosis 292, 294
self-discovery 42, 46
self-identification 292
self-improvement 222
self-injury 313
"self-narrating zoo exhibit" phenomenon 37, 215, 216
sensory disabilities/sensory sensitivities 3, 26, 30, 215, 222, 224
"severe" autism 172, 184, 309
severity scale, for diagnosis of autism 167, 175, 177, 179, 183–185, 187, 309
service provision 167, 168, 171, 177, 180, 183, 185, 188
Shattock, Paul 53
Shirley, Dame Stephanie "Steve" 280
Shoemaker, Clifford 96
Shore, Stephen 35, 214, 218, 230
Silberman, Steve 156
Sinclair, Jim 11, 23–37, 219, 307
Singer, Alison 148
Singer, Judy 2, 46, 213

single unified diagnosis, for autism 169, 171, 180
Skinner, B.F. 196
social communication disability, autism as 249
Social Communication Disorder (SCD) 180, 185
social communication domain, in the DSM-5 176, 181, 183, 185, 186
social conformism 300
social environment 7, 10, 313
social model, of disability 7, 11, 32, 125, 212, 217, 270, 272
social networks 11
social norms 7
social oppression 4
Something About Us 278
special education 32, 180, 184
"special interests" 175, 183, 185
standpoint epistemology v, 290
State Hospitals/State Schools (US) 79, 81
stereotypes 84, 125, 128
Stewart, Dr. Catriona 281
stigmatization viii, 2, 158, 170, 186, 313
 and AASPIRE, 135, 140, 141
 and Neurodiversity.com, 94, 98, 99
"stimming" 6, 8, 175, 185
strengths, of autism viii, 8, 9, 114, 184, 189, 307, 309
support needs 15, 170, 199, 308–313
Sykes v. Bayer 95
"symptoms", of autism 1, 7, 176, 184, 185, 312
synesthesia 147

T

Talking About Curing Autism (TACA) ix
Tammet, Daniel 223
thematic analysis viii
therapy ix, 7, 60, 230, 307
Thompson, Vilissa 244
Tisoncik, Laura 83, 90, 289, 295
torture, in name of treatment 195–207, 311
TPGA (Thinking Person's Guide to Autism, The) 13, 155, 156, 158, 162, 163, 165, 307
tragedy paradigm, autism-as- 26–28, 30, 33, 34, 91, 229
transgender 204, 292
"trapped" paradigm 28, 35
trauma
  autism as response to 188
  parental, 33
  suffered by autistic people, 43, 134, 195, 198, 203, 212, 222
  suffered by Judge Rotenberg Center employee, 202
treatment
  autism 74, 96, 99
  psychiatric, 54, 201
  torture as, 195–207, 311
Treehouse 278
Trump, Donald 238
twin studies 296

U

United Methodist Church 95
universal rights principles 4

V

Vaccine Court Chronicles 96
vaccines, and autism ix, 7, 8, 54, 155, 158, 223, 238, 239
  and Neurodiversity.com, 93, 94, 96–98
Ventura33 story website 123, 127–129
Vivian, Amanda 174

W

Wakefield, Andrew 54, 238
weaknesses, in autism viii, 7, 245, 293, 309
"walking while autistic" 70
Western biomedical model 8
"We've Only Just Begun" principle 250–251
"What Are the Stakes?" 180–181
WHO (World Health Organization) 229
Why I Dislike Person First Language 37, 219
Williams, Donna 214, 223
  and history of neurodiversity, 23, 24, 27, 28, 29–31, 32, 33, 37
Wing, Lorna 31, 55
workplace bullying 70
World Autism Day 151, 239
World Wide Web 42, 68, 214
Worlds of Autism 137, 139
Wright, Suzanne and Bob 227–230, 278

Y

"You Define You" principle 248

Printed in Great Britain
by Amazon